PRAISE FOR *FARMS WITH A FUTURE*

"Looking to get rich? Billionaire investor Jim Rogers says dump stocks and become a farmer! Or perhaps you simply long to nurture the earth and its inhabitants while feeding yourself and others at the same time. Regardless of why you're being called to farm, your journey begins here. In *Farms with a Future,* Rebecca Thistlethwaite shares a treasure trove of lessons, stories, and ideas from sustainable farmers who have already traded in the cubicle for the chicken coop. With Rebecca's help you'll be living your farm dream in no time!"

—**Tim Young**, author of *Poisoned Soil* and *The Accidental Farmers*

"What is so great about Rebecca Thistlethwaite's new book, *Farms with a Future,* is that it is not a generalized treatment about how to succeed in the new farm and food revolution, but a detailed and complete description of how some fourteen farms of various kinds have done it. This book does what books are supposed to do, that is, it gives the reader on-the-ground experience that would otherwise take years to gain."

—**Gene Logsdon**, author of *The Contrary Farmer* and *Small-Scale Grain Raising*

"*Farms with a Future* is an important book. If you are new to farming, read it now. If you've been in the business for a while, read it every year to keep yourself on track."

—**Shannon Hayes**, author of *Long Way on a Little* and *Radical Homemakers*

"*Farms with a Future* should be added to the reading list of anyone getting into farming. Easy to read and full of practical information, it will help those entering agriculture avoid many common pitfalls."

—**Nicolette Hahn Niman**, author of *Righteous Porkchop: Finding a Life and Good Food Beyond Factory Farms*

"Beginning farmers face so many challenges accessing the capital they need to get started or to grow their operations. But those who heed the excellent advice in *Farms with a Future* will be much more successful in their interactions with banks and other lenders."

—**Elizabeth Ü**, executive director of Finance for Food and author of *Raising Dough: A Complete Guide to Financing a Socially Responsible Food Business*

"*Farms with a Future* should be handed out to every budding young farmer in America. Thistlethwaite guides us through the daunting process of creating a viable small farm by sharing hard-earned lessons of her own, and from experienced farmers she met while researching this book. Indispensable!"

—**Novella Carpenter**, author of *Farm City: The Education of an Urban Farmer*

Creating and Growing a
Sustainable Farm Business

REBECCA THISTLETHWAITE

Foreword by Richard Wiswall

Chelsea Green Publishing
White River Junction, Vermont

Project Manager: Patricia Stone
Project Editor: Ben Watson
Copy Editor: Eileen Clawson
Proofreader: Helen Walden
Indexer: Linda Hallinger
Designer: Melissa Jacobson

Printed in the United States of America.
First printing November, 2012.
10 9 8 7 6 5 4 3 2 1   12 13 14 15 16

Our Commitment to Green Publishing

Chelsea Green sees publishing as a tool for cultural change and ecological stewardship. We strive to align our book manufacturing practices with our editorial mission and to reduce the impact of our business enterprise in the environment. We print our books and catalogs on chlorine-free recycled paper, using vegetable-based inks whenever possible. This book may cost slightly more because it was printed on paper that contains recycled fiber, and we hope you'll agree that it's worth it. Chelsea Green is a member of the Green Press Initiative (www.greenpressinitiative.org), a nonprofit coalition of publishers, manufacturers, and authors working to protect the world's endangered forests and conserve natural resources. *Farms with a Future* was printed on FSC®-certified paper supplied by Maple Press that contains at least 30% postconsumer recycled fiber.

Library of Congress Cataloging-in-Publication Data

Thistlethwaite, Rebecca.
Farms with a future : creating and growing a sustainable farm business / Rebecca Thistlethwaite.
p. cm.
Creating and growing a sustainable farm business
Includes bibliographical references and index.
ISBN 978-1-60358-438-8 (pbk.) — ISBN 978-1-60358-439-5 (ebook)
1. Sustainable agriculture—United States. 2. Sustainable agriculture—United States—Case studies. 3. New agricultural enterprises—United States. 4. New agricultural enterprises—United States—Case studies. I. Title. II. Title: Creating and growing a sustainable farm business.

S441.T433 2013
338.1—dc23

2012032205

Chelsea Green Publishing
85 North Main Street, Suite 120
White River Junction, VT 05001
(802) 295-6300
www.chelseagreen.com

Contents

Foreword

STARTING A FARM is a daunting task. Keeping one running is no less a feat. In fact, as occupations go, farming is one of the more challenging ways to make a living. And yet people are attracted to farming, and many succeed at it. You probably picked this book because you might be thinking of dipping in your big toe to test the farming waters, or maybe you are already in knee deep and want some guidance. In either case this book is for you.

Of the myriad of resources available for starting and running a farm, many are good sources of information for particular aspects of managing a farm—such as soil science, fertility, crop raising, animal care, and marketing, to name a few. Rarely does a book effectively encompass all the facets of the whole farm and how to map it out in a clear and concise fashion. Rebecca's book, *Farms with a Future*, does precisely this and tackles the crucial issues needed to start a farm and keep it thriving.

I met Rebecca a few years ago at a conference that she was organizing to help farmers with the business side of farming. As a farmer herself she intuitively knows the joys and pitfalls of a farming life. She is honest about the realities. *Lots of hard work*—check. *Learning curve is very steep*—check. *Things get hectic*— check. *Changing markets can be frustrating*—check. *Nature deals out some surprises*—check. But despite the potential mishaps, Rebecca enthusiastically endorses farming as a viable, meaningful, and profound occupation. *Working outdoors, being intimately involved with nature, taking joy in raising plants and animals, interacting with people through an honest business, adhering to green values, being autonomous, reaping the fruits of your labor*—checkmate.

"I want you to farm" proclaims Rebecca, a testament to her positive stance on farming. Beginning farmers rarely hear that encouragement, especially when it is backed up by a road map that leads to farm success. *Farms with a Future* covers all the bases, presented in an easy-to-read style. Where else will you find someone advocating that farmers not only keep an eye toward safety but also pay heed to their own physical as well as financial fitness? Rebecca includes profiles of a wide range of different types of farms, inspiring and enlightening stories of farmers across the United States and the issues that confront them. All farmers interviewed offer their advice for beginning farmers and keys to success, which make for some very valuable firsthand coaching. Some recommendations that farmers share have similar themes: for example, *"Start small before going big-time into a farming venture."* Or *"Don't try to do everything at once,"* and *"Expect hard work, but be smart about what you do."*

I was a beginning farmer once, but after three decades of farming, I'm getting closer to the "old-timer" end of the continuum. I remember the energy and excitement of starting up and trying new ventures, the joys and successes, and, yes, the failures (sometimes lots of them). And I was guilty of the "trying to do everything all at once." But as with any passion you have, the goal you want to reach propels you forward through thick and thin.

While there is no one "model" farm, there are common traits that successful farms share. This book offers up a blueprint for that success. Enjoy, and happy farming.

Richard Wiswall
Cate Farm
East Montpelier, Vermont

Acknowledgments

THIS BOOK NEVER would have come together had my family not made the decision in August of 2010 to take a break from farming. On that fateful day, one of the rare vacation days we ever took, while lying in our swimsuits next to a fantastic swimming hole, my husband and I watched our daughter play gleefully in the water. We hadn't had a family day like this in years, we thought to ourselves. In that instant, we decided to quit our jobs, end our farming business, sell everything we owned, and travel the country. As our plans took shape, we endeavored to add a research component to our itinerant travels—visiting farms that we had admired from afar or that other farmers spoke highly of. It is to those farmers we visited and interviewed for this book, and the thousands of other ones we did not get to, that we dedicate this book. The farmers and ranchers who are working to balance their own economic viability while providing good jobs, benefits to their communities, and reducing their environmental footprint, all while nourishing people with tasty, clean food—these are the farms with a future.

A special thanks, in alphabetical order, to the farms that hosted our family or participated in phone interviews: Amaltheia Organic Dairy of Montana, Big Bluff Ranch of California, Bluebird Grain Farms of Washington, Butterworks Farm of Vermont, Claddagh Farms & Cookery School of Maine, Crown S Ranch of Washington, Dey Dey's Best Beef of California, Fair Food Farm of Vermont, Green Gate Farms of Texas, Hoch Orchard & Gardens of Minnesota, Marilee Foster's Farm of New York, Massa Organics of California, Misty Brook Farm of Massachusetts, Nature's Harmony Farm of Georgia, Relly Bub Farm of Vermont, Serendipity Organic Farms of California, Shady Grove Ranch of Texas, Soul Food Farm of California, and Vanguard Ranch of Virginia. Although not all of these farms were profiled in this book, we learned so much from their generosity of time, food, and farming wisdom.

The content for the book you have in your hands comes from things I have learned along my bumpy path in farming and the farmers listed above. It also comes from the wisdom of too many books to mention, farming websites, and the amazing outside reviewers that worked with me on each chapter. I wanted to strengthen the substance of my book by engaging additional experts in their field —meaning they don't all have PhDs but years of practice. I am a practitioner at heart, so academics means less to me: it is what you do with that knowledge that counts. Starting at the beginning, chapter 1 was reviewed by Lindsey Lusher Shute, director of the National Young Farmers' Coalition. Organic produce marketing consultant Dina Izzo reviewed chapter 2. Steven Schwartz, founder and former executive director of the California FarmLink organization, reviewed chapter 3, and chapter 4 was reviewed by Elizabeth Ü, another Chelsea Green author and founder of Finance for Food. Chapter 5 was reviewed by one of my former UC–Davis professors of agricultural development, Dr. Paul Marcotte. My husband Jim Dunlop reviewed chapter 6, being a fanatic of creative farm infrastructure, good fencing, and buying used equipment. Chapter 7 was reviewed by a former mentor of mine and long-time director of the UC–Davis Student Farm, Mark Van Horn. Carrie Oliver, founder of the Artisan Beef Institute and Oliver Ranch Company, purveyors of fine dry-aged beef from respected ranches around the country, reviewed chapter 9. Chapter 11 was reviewed by Matthew Martin of Pyramid Farms, a 30-year veteran of organic farming and a fantastic financial manager, and Ron Strochlic, social justice consultant and former director of the California Institute for Rural Studies, reviewed chapter 12. Finally, chapter 13 was reviewed by Washington State Department of Agriculture's Education and Outreach coordinator, Sue Davis. Only chapters 8 and 10 did not get outside review, although I feel quite confident about their substance.

Introduction

If you aspire to farm someday, if you have just begun, or if you have been farming for years but want to create a more sustainable and economically viable model for yourself, then this book is for you. Regardless of your age or experience, there are some things you could learn that will make life a little easier for you and perhaps avoid a few pitfalls. I know—I have fallen into every one of them. You see, I am a farmer, one taking a break for a year to learn from other farmers across the country, to understand exactly what combination of resources, luck, and grit built their businesses and how they plan to continue into the future. My family and I observed, photographed, interviewed, and worked alongside dozens of farmers, large scale to small, to get a sense of the Right Stuff of Farming. Will all these farmers and ranchers be successful in every way? Absolutely not. They will have struggles, challenges they won't be able to overcome, mistakes they are not too keen to reveal. But they will also share innovations, foresight, creativity, and, most often, an unlimited well of perseverance.

A farmer mentor of mine once told me and a group of farmers-in-training that to be successful in farming you had to have the "Three Ts" and the "Three Ps." In Spanish those were *tiempo, tenacidad, y trabajo,* along with a generous portion of *pasión, perseverancia, y paciencia.* Translated into English, that is "time, tenacity, and hardwork" along with "passion, perseverance, and patience." When I saw this farmer recently, I told him the only thing I think his success list forgot to mention was **creativity**. With all the challenges you will face as a farmer, from competition to lack of capital, the ability to think outside the box will likely be your greatest asset. If you lack creativity now, this book will help you cultivate it.

Just what is a "Farm with a Future"? These businesses are run by farmers who are embracing their inner entrepreneur, who are not afraid of such things as marketing and sharing the pride they have in what they produce. These are also farmers who are building in systems, efficiencies, and.cost savings so they don't have to just keep raising their prices every year. They are shying away from huge debt loads and instead figuring out how to build their businesses patiently over time, using earned income or creative arrangements with their community of customers. They are farmers who are figuring out how to harness natural processes and how to rely less on purchased inputs, and who are careful not to degrade the natural resources that their farms depend on, such as clean water and living soils. They are farmers who treat their employees and volunteers like family and create community around their farms.

Many of these farmers are actively trying to earn a greater portion of the consumer dollar and go from being *price takers* to being *price makers*. They are differentiating their products in the marketplace, producing tastier or more-nutrient-dense foods, and communicating their values to would-be consumers. Not all of them direct-market their products, but even those who wholesale food are creating awareness and value around their brand. In short, these farms have a future because they are focusing on holistic sustainability, or what some call "the triple bottom line": economic viability, social justice, and ecological soundness. Without any one of these legs, the foundation of the business will not be sturdy enough to weather the ups and downs of farming.

This book does not idealize or romanticize farming. If you are not prepared for some serious hard work, inclement weather, dirt lodged in every crevice of your body, and being so dog-tired that you fall into your easy chair at night and don't wake up until the next morning, then you might look into another vocation.

Farming is not something you go into because you hate people, dropped out of school, or are unskilled. You will need other people to do this well, you will need to have a passion for lifelong learning, and you will need some skills. Of course, you can learn many of these skills on the job, but I recommend getting some training before you start your own farming enterprise or branch out into something new. Want to change from selling fluid milk to making cheese? This may seem like an obvious next step, but before you make it try taking a cheese class, reading every cheese book you can find, and maybe even volunteering a few hours a week with a nearby cheesemaker who is open to sharing the craft.

Sometimes you have to swallow your pride in being an "experienced farmer" and become an inexperienced one in a new field of learning. Apprenticeships and internships are available all over the country on farms of all shapes and sizes. You can also find paid farm work, from shearing sheep to driving a truck to market, fixing fence to selling veggies at the weekend farmers' market. Insert yourself into agriculture at many different points along the food chain so you will have a better understanding of all the links, who the players are, and the challenges and opportunities at each level.

My personal farming journey was a long and circuitous one, and I'm not done learning. I took a carpentry class a few years back and recently finished up an oxen-driving course—just a few more tools to add to my diversified toolbox. I am also going through holistic management training as I write this book, trying to understand better how good goal setting and decision making will contribute toward a more positive farming future for me and my family.

To get to the point at which I felt comfortable running my own farming business, I apprenticed on several farms, worked in food retail, conducted agricultural research, taught other beginning farmers, and eventually started my own farm with my spouse. Call me cautious, but I spent eight years thinking about starting a farm before I took the plunge. Perhaps you grew up on a farm and have been immersed in it your whole life. You still might need some training, especially if you plan to do something different from your parents' farming model. As several multigeneration farmers have told me, don't just copy the same things your parents did, because it likely won't fit with the marketplace of today. However, there certainly is a lot of wisdom and ways of doing things that could and should be emulated on a farm with lots of history. Learn all you can from your elders, particularly those who were privileged to grow up and learn self-sufficient, nonchemical approaches to agriculture that can serve us well today.

To start, I want to share with you a little about the evolution of our own farm, named TLC Ranch. You will see that as we scaled up over six years we also added and dropped certain enterprises. This is typical of many startups before they get to a level of maturation. Our final year was our most profitable because we finally had the right mix of enterprises at the right scale for our market and were selling at the right prices. However, it took many years of losing money or breaking even to get to that point. I often wish that we could have had all those issues of scale, right-size equipment, proper pricing, best sales venues, brand awareness, and so forth figured out earlier on so that we could have reached profitability sooner. I think having a book such as this one would have helped, along with more awareness of working business models and farmer mentors to bounce ideas off. But you know what they say about hindsight. We learned valuable lessons from running TLC Ranch and talking to other farmers around the country, and thus the genesis of this book.

Our farm started just like many others do. I had a full-time job, earning a decent salary, while my husband had some part-time gigs and began dabbling in farming. (He had farmed previously with another partner, so this was not new to him.) At first we began with eighty or so laying hens of diverse breeds so we could learn which types suited our production system and laid the quantity and quality of eggs we wanted. We also raised a few batches of seventy-five Cornish Cross broilers at a time in bottomless pens that got moved daily onto fresh grass, using a Robert Plamondon–style hoophouse chicken tractor. We built a simple chicken processing "shelter" and did on-farm whole-bird processing under the USDA-inspection exemption. The eggs and broiler chickens allowed us to get easily into a couple of farmers' markets that we staffed together on weekends. The money we earned went right back into buying more chicks, more feed, and building supplies for more layer and broiler shelters.

The second year we upped our layer numbers to around 300 and focused on one breed we liked best—the Production Red hybrid. We also increased our broiler numbers to around 1,000 birds over the season and raised 75 turkeys for the fall. We experimented with different movable housing designs and completely rebuilt our broiler houses using a lighter-weight metal conduit instead of wood framing to make them easier to move. We reseeded some pasture, paying a neighbor to do the tractor work for us, and experimented with different grains and grasses for our birds. Eggs and whole chickens moved easily at two farmers' markets, often selling out in the first hour. Again, all farm income went right back into the business, while my off-farm job paid the family's living expenses. We ate well and had time to garden, too (and take care of the new baby we had).

By the third year we increased our laying flock to 800 birds, produced around 10,000 broilers over the year, and tried raising a small flock of heritage turkeys for Thanksgiving sales. We dropped one farmers' market and added two others, bringing us to three weekly farmers' markets to attend, two of which were year-round. Time was starting to get tight, with Jim working full-time farming and me working full-time off the farm and part-time on the farm, and raising a toddler to boot. However, our third year we finally broke even on the farm expenses, yet we still couldn't come close to meeting the growing demand for our products. We also finally started raising prices as we realized that demand was far outstripping our supply.

Still in expansion mode, in our fourth year we enlarged our laying flock to 1,500 hens, raised our first batch of pigs, two Jersey steers, and a flock of 30 or so lambs but dropped broiler chickens and turkeys because an overzealous county health inspector misinterpreted the USDA exemption rule, which essentially disallowed us from selling into our main farmers' markets. We started out with a breeding pair of pigs that were given to us to settle a debt—no fancy breed but a solid set of "farmer hybrids" that produced great litters of piglets. We farrowed to finish around 20 pigs that year, and the pork sold easily at our farmers' markets. We also partnered with a nearby grassfed beef producer, selling around 10 steers' worth of his beef at our farmers' market booth, with him adding some of our pork to the markets he attended.

The fifth year we went a little crazy, raising 5,000 laying hens and growing out about 180 pigs that we either farrowed on-farm or purchased as weaners. We also added two full-time farm employees and one part-time farmers' market staff person.

In our sixth and final year of farming, we scaled our layer flock back to around 2,500 birds, upped our pig numbers to about 250, finished a flock of 30 sheep, and raised 5 Jersey steers that we bottle-fed. I won't give you exact numbers, but I will say that we grew by 430% in six years of business. For a time we were the largest pasture-raised producers of eggs and pork in California. But alas, being the biggest does not in and of itself mean success. We burned ourselves out and didn't have the quality of life we envisioned, renting overpriced farmland of poor quality in an area with underfunded schools and lots of crime.

So we closed that chapter of our lives to begin another. What we have learned this past year traveling the country has been extraordinary and inspiring. So much so that, by the time this book is published, we will likely be back at farming again, utilizing all of the knowledge and skills we have learned over our year of meeting and observing other farmers. We just can't keep our hands out of the dirt. . . .

Although the entire book is full of ideas that both beginning and more experienced farmers might find useful, chapter 1 is really catered toward the aspiring, just-starting-out farmer. The rest of the book is organized in the way that I hope you would approach your farming: starting off with the market in mind. I don't want you growing or raising a single thing until you have some concept of where you will sell it, to whom, for how much, and who your potential competition might be (chapter 2). Chapter 3 will help you consider the practicalities of owning versus renting land, where to find it, and how to secure it and a few innovative land-tenure models to think about. Then in chapter 4 we will discuss conventional debt financing, personal financing, investor financing, and other creative solutions to obtaining the capital you need over time.

Chapter 5 will be a lesson in planning, introducing you to the process of strategic planning, holistic goal-setting, and assessing your strengths and weaknesses. Chapter 6 will discuss what kind of equipment and infrastructure you might need to get started, depending on your growing system. Be prepared to spend a lot of time hunting for used equipment. If you only like new, shiny toys, then this is not the book for you. I want you to be profitable, not have a yard full of quickly depreciating, underutilized hunks of steel. Chapter 7 will go over irrigation and soil-management principles drawn from experts around the country and include references for where to get much more thorough information on these subjects. There are plenty of excellent agronomic books out there, and I won't attempt to go beyond scratching the surface on those topics. Chapter 8 will discuss various aspects of harvesting your crops or animals, postharvest handling, and further processing, with a focus on how to prepare your goods for market.

Chapter 9 will expand on where chapter 2 started by discussing how you will build a brand, methods of selling your foods, and communicating with your growing base of farm "cheerleaders." If you don't like people, I suggest you find somebody among your family or friends who loves them, because you will need a "coach" for your cheerleaders. Chapter 10 is the least sexy topic by far, but one of the most important. Slog through it anyway! You should have some understanding of the regulations that may affect you, even if it is only so you can steer clear of them. I will also cover some easy, painless methods of keeping records so April 15 doesn't come around with you scrambling to the nearest accountant with an overloaded shoebox of receipts in your arms, as well as record-keeping for the various third-party certifications you may choose. Chapter 11 will be equally glamorous, discussing some best practices for keeping your finances in order and your business afloat (as author and farmer Richard Wiswall calls it, "farming with a sharp pencil"). Chapter 12 will discuss the all-important topic of taking care of your people—yourself, your family, and any employees that you may have—how to treat everybody with dignity, respect, and equitable remuneration and take care of mental and physical health (yours included!). Finally, chapter 13 will share ideas for adding value to your farming operation, including innovative tourism opportunities, classes, and on-farm processing that could help even out cash flow, build community support, and engage more family members in your farming business.

Each chapter will also include one or two narratives of farms we visited that embody some of the best practices and creative ideas in agriculture—farms that we hope and expect to have a sustainable future. I used a completely unscientific process to find these farms: Some of them found me through my blog (www.honestmeat.com); some were recommended by other farmers ("You really should go see _________ down in _________"); and some I found by doing extensive Internet searches of producers who had won awards for their quality, their stewardship practices, or their innovations. I tried to cover the geographic diversity of this country but simply could not include every region or microhabitat of this sprawling, vast nation. I included farms near urban areas and those that are hours away from the nearest "civilization." I included farms in "red" states and "blue" states.

I also strived to cover a wide range of farm types and enterprise mixes, from grain production to dairies to row crops, orchards, and livestock. The majority of the farms profiled are organic farms, or at least employing mainly organic practices. Organic, ecologically based farming is the future of agriculture, as even the United Nations now agrees (UN Report, "Agroecology and the Right to Food," March 2011). Since this book is concerned with the future of agriculture, I thought I would profile forward-thinking, stewardship-minded farmers and ranchers to illustrate just how people are crafting and implementing their farming visions.

These farms are not perfect, nor are they stagnant: By the time you read this book, the farms may have added enterprises, dropped others, changed partners, found new markets, and figured out new ways to become more sustainable. That is one of their key features—they are flexible and adaptable—and I hope you will learn gems of wisdom from them, just as I did. Their perseverance has been an inspiration to me, giving me renewed hope that farming just might be a viable occupation for those willing to work extremely hard. Each farm narrative will describe how they got started, found and financed their land and infrastructure, chose their enterprises, and cultivated their markets, as well as how they approach sustainability and their advice for other farmers.

I want you to farm. I want you to take care of the land, the animals, the wildlife, the people while making a fair living for all of your effort. I want you to last—physically, mentally, and financially. I want to see a diversified, regionalized, ecologically based food system that supports farms and food businesses of many shapes and sizes, as well as the communities they reside in, but this will only come to pass if you have the right tools and support. I don't think we can all do it alone: There are too many pressures—regulatory, economic, and social—that are pitted squarely against the family farmers of this country. I am not so naïve as to think that sheer hard work will make you successful. We are all in this together: farmers, ranchers, processors, distributors, retailers, eaters, and regulators. But that is a topic for another book entirely. Onward. . . .

The Landscape of US Agriculture: A Condensed Version

The official statistics don't look good. Agriculture appears to be a dying art, yet we all eat. There are fewer farmers, with operations that are larger in size as farms consolidate, and those that are left are approaching their sixties (meaning that, as they retire, the future of their farmland will be in transition). Peel back a few layers, though, and something more hopeful emerges. There are more women as primary owner-operators, more ethnic minorities farming (except for African-American farmers, who have seen an almost complete extermination of their farming community), and ever more twenty-somethings getting started in farming. Farms are sprouting up on rooftops in New York City, on unused school fields, and in suburban backyards, and diversified operations are even growing out of monocropped commodity fields (see the movie *The Real Dirt on Farmer John* for one inspiring example in the heart of the Midwest). I won't bore you with the USDA agricultural statistics: You can find a treasure trove of interesting factoids on the USDA National Agricultural Statistics Service (NASS) website. Some commodity farmers are seeing record high farmgate prices, and for the first time in US history, the majority of our corn crop is being processed into fuel instead of being used as livestock feed or for human food products. However, as commodity prices go up, so do input costs, farmland values, and the American consumers' interest in capping or eliminating subsidies.

By the time this book is published, the legislative process known as the Farm Bill may have completely changed the long-standing tradition of directly subsidizing a small handful of commodity crops, as well as eliminated many of the conservation programs and reduced food assistance to the poor and needy (which actually makes up the bulk of Farm Bill spending). We have certainly met a lot of farmers who are not doing very well economically, experiencing a lethal combination of sagging sales with inflating input costs. However, we have met others who are managing to capitalize on Americans' renewed interest in home cooking, health and wellness, and buying locally whenever they can. To illustrate, a recent USDA Economic Research Service report estimated local food sales totaled $4.8 billion dollars in 2008 (direct to consumers or direct to restaurants/retailers), and the report predicted that figure would reach $7 billion in 2011. Because of the nature of data limitations, it is more likely that local food sales, including the value of bartering and trade, have a much higher economic value than this and are continuing to grow by leaps and bounds.

I think the future of US agriculture will be increasingly bifurcated, with more vertically integrated, large-scale farms focused on global markets and more diversified, small-scale, locally focused farms reconnecting Americans to where and how their food is grown. The "agriculture of the middle" will be an even harder place to reside as a producer, although more producer cooperatives and associations of smaller producers will help them play in the "big boys'" space without actually having to be Big Ag.

For the best historical analysis of US agriculture in the twentieth and twenty-first centuries, please read *The Unsettling of America* by Wendell Berry. Although it was written nearly 40 years ago, it all rings true today. The only changes since Mr. Berry wrote this seminal piece of work are that food production and food marketing are even more consolidated than when he wrote the book (Walmart, for example, now controls at least one-fifth of nationwide retail food sales) and US poverty and food insecurity are higher than ever before in the last 50 years (see www.worldhunger.org for more information).

CHAPTER 1

FOR THE BEGINNER

I DON'T CARE if you are 18 or 58, an urbanite or a country gal, you all have one thing in common: You have a desire to be a farmer but limited years of experience doing it—you're what some might refer to as a "greenhorn." You need resources and support, as well as mentors, friends, and family you can draw on for inspiration and know-how, or perhaps for startup capital and good old-fashioned extra hands. I have been a beginning farmer myself (and might even still be classified as one under USDA definitions), as well as worked alongside and conversed with hundreds of beginning farmers across the country to get ideas for this book. Here, in a nutshell, is what I have concluded from their collective wisdom:

1. Dabbling in farming is fine when you start—in fact, it is encouraged. But don't go too long without creating a **plan**.
2. Writing a formal **business plan** is a good mental exercise and will be instrumental if you want to apply for financing later on in your business evolution, but when starting a new business you can expect a lot of **change** in those first few years. Unless you intend to draft new editions of the plan with every twist and turn, your business plan may become obsolete immediately. Shoot for having a fairly ironed-out plan by year three. Make sure you include all members of your family in the planning process if they have a role or stake in the business or intend to in the future. More than likely, you will not be doing this alone. Communicate with all the likely stakeholders.
3. New businesses rarely make money in the first few years and often lose money while they are gearing up production and working out the kinks. Have a **survival plan** for those first few years of loss, and try not to borrow against your future profits because, if you do, you may never get to profitability when your cash flow is too tied up servicing debt. If you are continuing to lose money into your second or third year, you should do a thorough

analysis of your business model to understand how you can turn that around. Actually, do an end-of-the-year analysis every year until you start to turn a profit. A more in-depth discussion on financial management will come in later chapters.

4. **A farm is a business**. If you don't want to run a business, consider gardening or homesteading instead. You will have to understand such things as profit and loss, assets and liabilities, and supply and demand and do basic bookkeeping. If these things scare you, that's okay, but be willing to learn about them.
5. **A farmer need not be poor**. It is reasonable that you should be paid for your time and effort and that you make enough money to put away for a rainy day or retirement. You are your farm's greatest asset—make sure you protect that asset! So plan for profit right from the beginning.
6. Start with the **market in mind**. Spend time researching potential customers, market channels, food fads, and so on. Look for gaps in the local food market that you might fill. Begin thinking about how your products will be different, superior, and so forth—more on this subject in the next chapter.
7. A garden or homestead is one of the best places to **test your farming ideas**. So is working for other farmers. Start small, build your skills, learn what grows best in your soils and climate, and figure out what you actually *enjoy*. If you can't stand the behavior of a couple of sheep, you may not want to become a sheep farmer. If you hate bending over all day, you may not want to become a strawberry farmer. If baling and putting up hay turns you into an allergic mess of snot, you may not want to make hay for a living. Figure out what you like and excel at, and begin with those enterprises. You can always add or even delete enterprises later, but make sure you develop your farm around activities that bring you some joy or satisfaction.
8. If you wanted to open up a new restaurant, you would want to have some capital to do so. If you sought to build a new widget factory, you would want to have some capital to do so. If you want to start a farm, you might want **some capital** to do so. This may be hard if you are young and have a limited number of years working and saving money, or you are older and just barely getting by. However, you can start farming part-time while you work an off-farm job (indeed, 70% of American farmers do this), earning income that will support your basic living expenses and provide some capital for startup costs. Unlike many other businesses, you can also use

the power of **social capital** to start your farm: Many people enjoy helping farmers by lending money, sharing equipment, or even providing volunteer labor. Much more on this subject later.

9. Start thinking about the **scale of your business** right when you start. A great exercise to get you thinking about scale is to project how much income you would like to make in **years 5 and 10** of your business. Work backward from there to understand how much total revenue you need to make in those years to earn you that income. Then begin planning budgets from year 1 to year 10 to understand how you will have to scale the business to get to those numbers. A rudimentary example is charted below, which assumes that your goal is to make $45,000 in net profit by your tenth year of business:

YEAR	YEAR 1	YEAR 2	YEAR 3	YEAR 4	YEAR 5	YEAR 6	YEAR 7	YEAR 8	YEAR 9	**YEAR 10**
Total revenue	15,000	19,000	22,000	26,000	31,000	38,000	45,000	53,000	62,000	73,000
Costs	18,000	19,000	20,000	21,000	22,000	24,000	25,000	26,000	27,000	28,000
Net income	-3,000	0	2,000	5,000	9,000	14,000	20,000	27,000	35,000	**45,000**

What is a typical beginning farmer narrative? There is so much diversity in terms of age, gender, ethnicity, and geographical origins of today's beginning farmer that no beginning farmer is "typical." Many aren't even coming from agricultural backgrounds or a historical family farm anymore. However, I think the two things that unite all beginning farmers are their enthusiasm and their fresh ideas. Here are a couple of eager farmers getting started in the Green Mountain State of Vermont to get you thinking and hopefully inspired.

From Virtual to Reality

RELLY BUB FARM, WILMINGTON, VERMONT

• Nathan Winters •

Riding your bike across the country for five months shows a lot of tenacity, endurance, and patience. These are precisely the characteristics that a new farmer should possess. Nathan Winters came to the realization that he wanted to be a farmer while visiting farms of all shapes and sizes via his bicycle over the summer of 2009. His goals as a new food producer? To encourage people to break free of the industrial food model, enjoy and restore the art and value of cooking, share meals together, and buy food that is in line with their values and that is grown with compassion for animals, enhances our soils, keeps the interest of our future generations at heart, and provides optimal nutrition.

After Nathan returned from his big trip, he jumped into working on a diversified organic farm in Northern Vermont called Applecheek Farm. There he learned how to care for animals, dabbled in vegetable production, and got a closer look at the business side of running a farm. He also helped the farm's owners home in on their social media marketing, since he brought a background in software development and social media from his previous career. This skill has proven to be a major asset to the Relly Bub marketing plan as Nathan gains attention in the local community and actively promotes his farm foods in advance of his first commercial season.

Nathan found his farmland in Wilmington, Vermont, a small town situated halfway between the bigger towns of Bennington and Brattleboro, through a humble Craigslist wanted ad. In a short paragraph he explained that he was looking for a house with a small amount of land to farm. Immediately, a family with a gorgeous estate of hundreds of acres contacted him about a small cottage they had for rent. Eager to get started, Nathan decided to rent the place using a casual month-to-month agreement to begin with. "We keep the lines of communication open" is how Nathan described the new relationship with his landlord, a family that shares the same values as Nathan of ecological stewardship and organic gardening. In fact, his landlord Frieda's food garden was one of the most stunning, abundant, and healthy gardens I have ever seen, right out of a *Better Homes & Gardens* spread. That probably bodes well and suggests a landlord that will understand food production and support these new farmers. As Nathan grows and becomes more serious about his commercial farming endeavors, he will probably ask for a secured lease—a smart move for any new farmer.

Vermonters take care of each other, something that was evident when the brother of Nathan's new landlord came out with his tractor and immediately began tilling up nearly an acre of fallow land for Nathan. Frieda also brings Nathan large bags of grass clippings that he uses both in his compost bin and as mulch in the garden, and she also provided Nathan with many of the vegetable starts he needed in his first season. A large homestead garden is what Nathan essentially planted his first year, experimenting with a large diversity of vegetables and culinary herbs to see what grows well in the short but hot Vermont summer. Nathan also tilled up an equally large patch that will host next year's garden and planted it to buckwheat and oats for a cover crop and for his pastured broiler chickens to fertilize.

Raising 30 Freedom Ranger broilers in the first year gives Nathan a chance to hone his skills, and figure out how often to move the birds and the proper shelter/feeding/watering arrangement, but it is

Nathan inspecting his first-year market garden.

also not too daunting a number that he can't butcher the birds himself. He will have a chance to run the numbers, feel for demand in the community, and create the ideal scale next season. The same decision was made for the laying hen flock: Start small, experiment with building some low-cost shelters made from found materials, and figure out if eggs make economic sense before going hog wild (or chicken wild!). Nathan is also excited to add pastured pigs in the future, which was his favorite animal to raise on Applecheek Farm, but in year one he is focused on building his skills with organic vegetables, broilers, and layer chickens.

Future plans include building a self-serve roadside stand (Nathan lives on a well-trafficked road with an endless amount of tourists passing through), perhaps even with a "take what you need, pay what you can" philosophy. For Nathan it is important that everybody, regardless of income, has access to good, organic food. In addition to the farmstand, Nathan may attend the small summer farmers' market in Wilmington, which could surely use more fresh vegetables, and potentially partner with other farmstands in the region that might be looking for more local produce.

Luckily for Nathan, even though there is certainly the interest and income to support more local farmers, there are actually very few around the area where he lives, at least those that direct-market. There are also numerous restaurants and even ski areas that are likely looking for fresh, local ingredients. Nathan is particularly enamored with the CSA model to secure guaranteed support early on during the season as well as build transparency with his customers: He is already working to get that off the ground next year.

However Nathan decides to scale up, he insists that he doesn't want to have the kind of farm that consumes him and takes away his joy for living. Nathan wants to have time for himself and friends, his dog Chaya, his writing, and maybe even travel during the winter dormant season.

Eventually, Nathan would like to see the majority of his income derived from farming and enjoy a supplemental income through his writing during the winter. Nathan is nearing the completion of his book, in which he shares stories from his farm-to-farm journey across America along with his personal transformation into agriculture. At some point Nathan wants to settle down, have a family, and own his own piece of land. Thus far, the right partner has not come Nathan's way, and he believes Relly Bub Farm will serve as the perfect place and opportunity to meet someone who has a desire to live a life of simplicity.

To those ends Nathan is spending this year in deep-observation mode, soliciting advice, experimenting with different crops, varieties, irrigation methods, mulching and weed control, and cooking as much as he can from the garden. (He wants to provide cooking tips and recipes to his customers through his farm blog, so he better darned well know how to cook the stuff himself!) Luckily for Nathan, he is a very competent home cook, regularly churning out homemade pizzas, breads, quiches, canned goods and more.

So what about the cute name "Relly Bub Farm"? Well, Nathan loves his dog Chaya, and while working at Applecheek Farm last year, he would always be amused when Forrest, one of the young farm kids, would lay his head down on Nathan's dog and say, "Chaya, do you want a relly bub?" (instead of "belly rub"). Those words stuck in his head until it came time for him to think of a farm name. Not only does the farm name bring a smile to your face, it is memorable.

Nathan is a strong believer in using the Internet for networking, learning from others, marketing products, and building community. Despite barely having anything to sell to people yet, Relly Bub Farm has its own website, blog, Twitter feed, Facebook fan page, and Flickr photo album. It's not only about building customer anticipation, it's about transparency, says Nathan. He wants to share with people how he is creating his farm, show photos of the great things he's growing, and build enthusiasm in the community.

The makings of a future CSA box

NATHAN WINTERS

Creating a farm that is not only light on the earth but is actually restorative of the soil, biodiversity, and the soul is important to

NATHAN WINTERS

Laying hens prepping the next garden plot.

Nathan. As Nathan remarked, "We need to move past sustaining—why would we sustain the status quo? We need to take a more restorative approach to everything."

Nathan's advice for other new farmers?

- Don't be afraid to admit you don't know something/don't be afraid to ask questions.
- Be humble.
- Reach out to other farmers for advice, inspiration, borrowing equipment, etc.
- Take it slow, be patient, don't be in a rush.
- Be clear about your goals and vision.
- Don't work yourself to death (it's not cool to work 90 hours a week; don't lose your balance in life).

FARMER FITNESS

Along with getting a good dose of skills training, get your body ready for long hours and physical work. This may seem obvious, but just as you would physically train to be a firefighter, you should do the same if you want to farm for a living. Add aerobic conditioning to your life, at least 45 minutes a day, and include some weightlifting or resistance training into your regimen a few times a week. To decrease injury and manage the stresses that will inevitably come your way once you are farming, try yoga, tai chi, or some other form of slow stretching and meditation. You don't have to become a yogi or a marathoner; just begin to make a habit out of exercise and stress management. Learn how to lift heavy objects appropriately, the most ergonomic way to use tools, and how to sharpen and maintain those tools to work safely. A dull tool might lead to a hurt back or worse. You must sustain your body to have a sustainable farm.

If you want to farm for a living and not just as a hobby, then you need to get financially fit, too. Start a savings account, obtain a copy of your credit report and start fixing any problems that you see on it, establish a credit history if you have not done so already, and pay down your debt starting with the highest interest rates first. Make an appointment at your local Farm Services Agency office to find out what you need to start doing today to qualify for a farm loan in the future. Keep your day job, if you have one, or think about other ways to have your basic needs taken care of while you pour all your money into farming. (Live with your parents! Live in a tent! Find a spouse with money!) One caveat: What's the best way to make a small fortune in farming? Start with a large one! In general, farming is a lousy way to invest money, whether it is yours or someone else's. The financial returns are often small or nonexistent, but the other rewards will, I hope, outweigh this fact. You don't have to be rich to farm, even though there are many rich people who farm. You will need to be strategic and smart about your money because farming is a business, too. I will get into this subject in more depth in chapters 4 and 11.

Here is another beginning farmer couple, also in Vermont, who are developing a farming model that attempts to feed their immediate community in a socially conscious way. Their "low-profit" model, while certainly commendable, is also challenging as a startup business, which must have profits to reinvest in itself if they expect to grow and take care of their household's financial needs. However, I think they have many of the elements needed to create a successful and unique farming business model, which is why I have included them here.

A New Kind of Small-Town Grocer

FAIR FOOD FARM, EAST CALAIS, VERMONT

• Emily Curtis-Murphy and Matt Yetman •

Although they had numerous years of experience working on organic farms in Northern Vermont, Matt Yetman and Emily Curtis-Murphy were at first divided about the idea to start their own farming venture. Matt recognized the risks and enjoyed the stability of a weekly paycheck and not bringing work stress home with him. Emily knew they couldn't support a family on a couple of farm worker paychecks and wanted the freedom to craft her own farm in line with their values. So they began to hunt for land all over the Northeast. Pretty quickly they came to the realization that they were not going to be landowners, at least not yet. A little community near Montpelier, the capital of Vermont, created a new town plan that called for supporting agriculture and beginning farmers. That sounded like it just might be the place to get a farm off the ground.

Matt and Emily attended a Conservation Commission meeting in the town of Calais (pronounced "callus," like the ones you have on your hands) to let everyone know how eager they were to find land to rent. They learned of a couple with a beautiful, sprawling parcel on the ridge above town who were

Organic crop rotations with laying hens for bug control

EMILY CURTIS-MURPHY

Who says you can't have local food in winter?

keen to see their hayfields farmed with food crops instead. A written lease was signed that allows them to use about 45 acres of land, 10 of which are tillable and the rest will be used as pasture for animals. No cash is exchanged; instead the couple supplies farm-fresh food as their payment. They wisely negotiated a five-year lease and used a template they found online to craft the agreement. They figured five years would give them some security to build their business, but some flexibility as well.

At the same time that Matt and Emily were looking for land, they were trying to identify the optimum place to build a farmstand. They knew direct marketing was going to be their best chance to earn the most for their products, and they were intent on providing food for their local community. A fortuitous "For Rent" sign was found on a handsome old grain mill next to the creek in East Calais Village. The building has a lot of character and a microhydro system to boot, which helps offset some of their electricity costs. After installing a walk-in cooler, and doing a lot of cleanup, painting, and landscaping, the Fair Food Farm store was born.

Their first growing season was 2010. Although the hayfields had good fertility that first year, the weed pressure was exhausting. Still, Matt and Emily grew a lovely assortment of vegetables and raised some animals, too. The farm store's sales were good with the buzz of their opening and summer tourists and second-home vacationers coming into the area. In the spring and summer of 2011, the closure of the main artery coming into East Calais because of a bridge replacement project severely hampered their sales. When we visited, summertime sales were just picking back up again.

Running a new farm and a new store (and having two babies) all at once has been arduous for Matt and Emily, but their dogged determination has kept them going. Their little store is filled to the brim with organic vegetables grown by them or purchased from other local farmers, pastured meats that Matt and Emily raised (right now that consists of pork and broiler chickens) and some grassfed beef purchased nearby, along with local dairy products, grains, beans, maple syrup, and a few exotics such as Vermont-roasted coffee and some killer Fair Trade chocolate. They would like to add more nonperishable foods, but their cash flow limits the amount of inventory they can stock, especially if the item is going to sit on the shelf for a while. The store is open seven days a week and is mostly self-serve in the mornings, which is an amazing concept in itself. Could you imagine going into any grocery store, picking out what you want, adding up your total, and leaving your cash in the drawer? That level of trust is palpable.

The Fair Food Farm name was chosen because of the strong value that Matt and Emily hold of making good, organic food available to people of all income levels. As part of that mission, Fair Food Farm is organized as an L3C, otherwise known as a low-profit limited liability corporation. It combines the benefits of a traditional LLC (reduction in personal liability) with a stated social mission, something that potentially enhances the farm's ability to receive program-related investments (PRI) from foundations. Thus far, Matt and Emily have not received any PRIs, but they plan to solicit such support as they become more established. Foundations that care about rural food security and organic farming might see Fair Food Farm as a good potential model worth investing in.

Right from the beginning, Fair Food Farm offered a 15% discount on food purchases for anyone using food stamps. Last winter they leased an electronic benefits transfer (EBT) machine so they could also accept food stamps (not even the well-established food co-op in town accepts food stamps). Not many low-income families are taking advantage of this benefit; however, Emily expects that will change as more and more people hear about their store. Another wonderful benefit that Fair Food Farm provides are several subsidized CSA shares. A low-income family only pays 50% of the CSA share, Fair Food Farm fund-raises for 25% of the cost, and the Northeast Organic Farming Association of Vermont comes up with the other 25%.

This unique program organized by NOFA-VT assisted over 1,400 families to become CSA members in 2010 and undoubtedly increases their consumption of healthy fruits and vegetables. For customers who may have more time on their hands than money, Fair Food Farm offers a number of work-trade slots each year. Essentially, for each hour that a person works, either on the farm or staffing the store, he or she receives $10 of food credit. I met two of their work traders who were both unemployed and were extremely grateful that they could still get high-quality organic food into their diet via this exchange. It has also provided a source of labor for this small farm, which can't afford to hire employees yet.

This year Matt and Emily are growing a wide range of vegetables, storage crops for the winter, around 50 pigs, 900 or so broiler chickens, 50 turkeys, and a flock of laying hens to supply eggs for their store. They are building a new hoophouse to extend their vegetable season, thanks to a cost-share program via the USDA Natural Resources Conservation Service (NRCS). Planning the correct volume for their store has proven difficult, since sales ebb and flow with the seasons, roads wash out, and word slowly spreads about their great little store. One way they have ensured a small, steady customer base is through their unique storewide CSA. Customers pay up front, essentially providing a credit that they deduct from over the course of the year. Then they can take any foods they want from the store. Currently, about 30 customers are utilizing this program, and enthusiasm for it is increasing.

Signage at the farm store tells the truth.

Fair Food Farm also offers a traditional box CSA, which is a great way to get people to eat with the seasons and try different vegetables that the farm can grow. However, there are already quite a few CSA farms operating in the area, so the store CSA might be a way they can better capitalize on their unique situation of running both a farm and a store.

Emily was especially savvy and creative about the financing of their startup farm and store. The first round of financing involved a lot of begging and borrowing from family and friends. The second round included a lengthy, paperwork-heavy process of applying for a loan through the Vermont Agricultural Credit Corporation. The third round involved connecting with a "Slow Money" private investor, who loaned just enough to get Matt and Emily to their projected budget for that first year (including both operational costs and some long-term assets, such as a tractor). However, this year they were not able to reapply to the credit corporation for another operating loan because all of their collateral assets are already promised to the first loan. Consequently, they have relied on small customer loans, and the cash flow has been extremely tight. Emily likens their bank account to a "sieve." Cash flow is often an enormous struggle for beginning farmers when they are trying to get to the right scale to support the fixed and variable costs of production. Emily has been spending a considerable amount of time doing business planning and cash flow projections to get on top of their financial situation.

Fortunately, Vermont has a very food-focused culture, with both highly educated consumers and a plethora of direct-market farmers. However, Fair Food Farm has had a hard time penetrating the local markets because many potential customers (such as restaurants) already have long-standing relationships with existing farmers or, in the case of the farmers' markets, there's a waiting list. And as mentioned before, CSA farms also abound in the area. Matt and Emily's plan—to increase their local customer base by reaching out to families of diverse incomes—has been less successful than they had hoped. This has forced the couple to start expanding their geographic reach and consider markets as far away as Boston. Luckily, they have family near there that can help watch their kids while they attend farmers' markets.

In the meantime, they are doing all they can with a limited budget to convince locals in their rural community to shop regularly at their store. Emily makes a point of periodically checking the prices at the nearest chain grocery stores in Montpelier, then returning to her store to make sure her prices are below those of the chains. Her commitment to making their food affordable is admirable, but will it sustain their business over the long term? My hope is that the community will realize this great asset they have for creating local food security and make it a habit to shop at the Fair Food Farm store.

Emily's advice for other new farmers?

- Start small and *slow*.
- Don't take on too many new enterprises at once, especially ones that you don't have experience with.
- Start dealing with the weed seed bank as early as possible. Get your soils ready in the fall for the following season.

TAKE-HOME MESSAGES

- Take stock of your personal finances when you are getting started. How is your credit score? (Print up a free annual credit report, and read it thoroughly.) Are you starting with a significant amount of debt? How is that debt structured—will you have decades to pay it back or only a few years? Will you be able to take on farm-related debt in addition to your personal debt and still have a reasonable debt-to-income ratio (under 35% or so)? I will explain this ratio later in chapter 11.

- Farming requires capital and cash flow. Understand your annual cash flow requirements and the seasonality of revenues versus costs. (For example, you still have to feed the animals in the winter when they might not be producing, or you have to buy seeds in February, even though you may not have income coming in to pay for those seeds.) Create a budget projection for the entire year, which will help you foresee some of those cash flow shortfalls.
- Don't plant too much or start with too many animals before you have the cash flow to support them. Every single farmer I interviewed said the same exact thing: *Start small.*
- Feel free to experiment, but don't grow the wrong things for your soil, climate, and land base.
- Don't produce what everybody else is producing; have a plan for differentiating yourself.
- Don't grow without an identified market or committed buyer.
- If you don't have the work ethic or time to see things through harvest, then don't waste your resources on it. I can't tell you how many new farmers I have seen till under poorly performing crops or liquidate flocks/herds of animals because they are growing poorly or are unhealthy due to neglect. You don't have money to give away like that.
- Don't enter farming with either a romantic or a sexy image in your mind, because farming is neither.
- Don't throw caution and safety practices to the wind; don't injure yourself early on.
- Continue to make sleep, personal time, and family/friend time a habit from the beginning. (Don't fall into the "I'll sleep when I'm dead" routine.)

About five years into our last farming business, I had the chance to meet with a professional business consultant for two hours of free consultation. He quickly took stock of our business's strengths and weaknesses, as well as our financial performance. When it came time to give me some preliminary advice, he asked, "Are you ready to give up this hobby and make it a real business?" For a moment I was dumbfounded: I thought we *were* running a business. I mean, we filed our business tax forms each year, had an employee identification number (EIN), and even paid a bookkeeper to keep our finances in order. From what I could tell, we were even making a profit. I pressed the consultant for more information.

I can't remember exactly how he stated it, but he told us that we needed to stop flitting about with different enterprises, starting new ones and stopping others, and instead we should focus on the few that were stable and profitable. We needed to fine-tune their production, really learn to cut costs and maximize efficiency, market ourselves more effectively, and do higher-level accounting to truly understand our profitability. Had we ever created a strategic plan, he asked? (Not really, just a five-page business plan that we hadn't updated in several years.) He also suggested we make time for organized family business meetings instead of the informal day-to-day conversations we often had, which eventually consumed me and my husband's every interaction with each other.

Why I hadn't sought out the help of a professional earlier on in our business is beyond me. We farmers tend to be an independent, do-it-yourself lot of people. But there is so much good help out there, much of it available for free. One of the best places to start is your local Small Business Development Center. Next time our family starts another farming business, we will be taking full advantage of our nearest SBDC and the mentors they provide. I expect this book will provide you with some of those resources to take your farm to the next level. If you want to stay a hobbyist for the rest of your life, that is absolutely fine. My hope, though, is that some of you can make farming a career and a financially viable pursuit that is also ecologically and socially sustainable. Our nation's food security and our planet could surely use a bunch of you.

CHAPTER 2

IDENTIFYING YOUR MARKET NICHE

Say you grow a field of straight, crunchy, sweet carrots; then harvest, wash, bunch, and pack them all in wax-covered boxes. Next, you pick up the phone and call all the grocery stores in your town. They all give you a lukewarm reception and offer to pay you whatever the market rate is, currently $0.75 a bunch. Doing a little back-of-the-envelope math, you figure you have to get paid at least $1.00 a bunch to break even. Searching for other options, you call a handful of local gourmet restaurants to see if they would be interested in buying carrots. They are, but they only go through a box of 24 bunches each week, so that helps you move only four boxes of carrots when you have a pallet of them waiting in your cooler. Desperate, you start to call markets even farther away and leave messages with a bunch of distributors.

You can't get into the farmers' market with one crop, nor can you entice everyday consumers to stop at a roadside stand stocked with only carrots. Eventually, you wind up finding a buyer willing to pay $0.90 a bunch and you have to drive an hour to drop off the pallet. Factoring in the mileage on your vehicle, you would have been better off selling locally for less money. In the end, you lose money, along with a chunk of confidence, and scale back your farming plans for the future based on this one experience.

None of this needed to happen! Take a different path, one that involves some upfront homework but that will enable you to sustain your business over the long term. In short, before you take on any farming enterprise, do some market research.

MARKET RESEARCH

You don't need to be an economist or hold an MBA to do good market research. It is fairly straightforward: simply ask and get answers for the following questions:

1. ***Who* is your potential customer base?** This could include individuals, distributors, stores, restaurants, institutional food services, food processors, and so on. Get demographic data on the individual customers (income, ethnicity, and family size are key variables) and business data on the business customers (gross revenues, food budgets, other suppliers they buy from, reputation, who their typical customer are, etc.). Give a handful of prospective business customers a call to find out what they might be looking for. Don't rely on their word as a contract to buy; however, it may give you a sense of *market potential.*
2. ***Where* is your potential customer base?** Will you be selling off the farm only (this may be required for certain raw-milk producers, for U-pick, farmstands, etc.), in your local community, regionally, across the state, around the country, or for export? Will you be able to satisfy the logistical issues involved in getting food into these different locations (washing, sorting, packing, freezing, drying, food safety protocols, etc.)? As the distance increases, so often do the complexity and the number of middlemen (handlers) earning money off your product (meaning you will be earning less of the consumer dollar).
3. ***What* are the current gaps in these potential markets?** Are there specific products that seem to be in short supply or simply missing? Or are these markets missing high-quality or values-based products? There may be a good reason these foods are missing or in short supply. Try to understand why that may be. You don't want to assume there will be demand for a product just because nobody else is doing it. Other farmers may not be growing that crop or animal for a good reason (climate, soils, daylength, disease pressure, poor economic performance, logistics, etc.).

   However, just because the market demand for a product might currently be a little soft does not mean that can't change over time. Consumers are growing in their education level about the food system and their desire for healthy, local food. A farmers' market we used to attend seemed to be full of young college students and vegetarians, meaning our sales of meat were somewhat disappointing. After several years of nonstop education and mind-numbing Q-and-A sessions, many of these vegetarians started adding local humanely raised meats to their diet, and our products were their first choice because of our commitment to transparency and education. They became some of our best customers and strongest allies, but it took time. As a farmer or rancher, you need to be constantly aware of what the new trends or opportunities might be (make friends with at least one chef and one retail buyer). If you are an early adopter, you may be able to capture the market share and be considered the "go to" source of a food.

4. ***What* are others charging for similar products?** Although I believe strongly that you should price your products based on *your specific costs* of production, you should have an understanding of the pricing of other similar products in your marketplace. If farmers' markets will be a venue for you, what are other vendors with crops or animal products similar to yours charging? Are your products superior to theirs, such that you could charge higher prices? If wholesaling will be part of your sales mix, look at wholesale price lists from similar farmers or ask your buyer to furnish you copies from other vendors. Ask chefs what they are willing to pay and under what scenarios they would be willing to pay a premium (certified organic, humane animal care, unique flavor, heirloom variety, picked that morning, etc.). It may be possible, at least for the time being, that what you want to grow and the way in which you want to grow it will not command the price you need to raise it.

   As an illustration, while attending farmers' markets in some rural and economically marginalized areas this year, I have seen organic eggs sell for as low as $3 a dozen. Under the current organic feed prices (which have been incredibly high for the past several years), I know that it costs between $2 and $3 a dozen in feed costs alone to produce these eggs. If the market will only bear a price close to $3 a dozen, it may not make sense for you to produce eggs at this time, or to produce them for this particular market segment, because you will not be profitable. However, test the market first (on a very small scale, before making any large investments) before you make the assumption that your local market will not bear the fair price—you may have a higher-quality product that consumers will recognize and reward.

If you will have to travel farther to get to higher-priced markets, will the transport costs negate any potential price gains? (I will discuss prices in more detail later in this chapter and in chapter 11. It is such an important topic that it warrants a lot of discussion.)

To expand on the subject of market research, let us discuss some of the nationwide food purchasing and consumption trends. This will be essential for good planning. With the economic downturn of the last several years, unemployment and poverty have risen, driving American consumers to look for more value in their food purchases. Coupon redemption is up 25% from 2009 to 2010 (Compass Natural, 2010), eating out is down, and Walmart is the #1 grocery retailer in the nation (SupermarketNews.com).

However, organic food and beverage sales are up 7.7% from 2009 to 2010, generating around $26.7 billion in sales for the organic "industry" (Organic Trade Association [OTA], 2011). This now represents over 4% of the US food market. There are more retailers than ever selling a wider variety of organic food. Even nonfood organic sales are up, reaching almost $2 billion in sales of such items as organic fiber, beauty care products, and supplements (OTA, 2011). A broader range of Americans are adding organic food to their grocery list: Indeed, the latest research shows 78% of Americans say they purchase organic food at least sometimes (OTA, 2011).

Although food trends change rapidly, especially the list of "hip" ingredients, there are some interesting changes afoot that are worth paying attention to. Can you position your farming enterprises to take advantage of any of these trends? Will you be an early adopter or come in later once a concept has been "proven"? Here is a list of some noteworthy trends, adapted from a nutritionunplugged.com blog post dated 5/8/2011 and from Steven Hoffman of Compass Natural, in no particular order:

1. **Changing Demographics**—Food will be more demographically directed, targeting flavors, foods, and messages to different generations and ethnicities.
2. **Cooking at Home**—Home cooking is on a slight upswing after being on the decline for a few decades. However, many Americans are inexperienced home cooks or lack confidence in their cooking skills; indeed, 28% *don't* cook because of their perceived lack of skills (Impulse Research, 2011). They are still looking for

ways to cut time and make dishes more "foolproof." Think frozen, precut ingredients, as well as heat-and-eat dishes.

3. **A Return to "Americana"**—There's been an upswing in anything that is or perceived to be local or farm raised or has traditional "American" flavors. Regional American cuisines are increasingly popular.
4. **More Natural, Less Processed**—Foods that have fewer ingredients, no preservatives, no "chemical-sounding" ingredients, or more whole grains are becoming more popular. "Natural" appeals more to consumers than "organic" does now, despite the fact that the term is unregulated and sometimes used in a misleading way.
5. **Rise in "Functional Foods" and Eating for Nutrients**—Consumers are looking for food fortified with or naturally high in protein, omega-3 fatty acids, antioxidants, fiber, and so on.
6. **Desserts and Retro Flavors Making a Comeback**—Despite our obsession with weight, we are splurging on desserts at least two times a week. Caramel, butterscotch, malt, and vanilla are making a comeback. Homemade ice cream, popsicles, and cupcakes are hot, but there is also more interest in sugar-free and gluten-free desserts.
7. **Return to Three Squares a Day**—Eating breakfast at home is on the upswing, as are to-go breakfast foods such as breakfast burritos. Ethnic and minisandwiches are also popular for lunches.
8. **More Entertaining at Home**—Folks are entertaining more at home and less by going out.
9. **Value Plus**—Companies that can communicate both good price value and other benefits (health, environmental, social) will continue to grow in this tough market.
10. **Sustainable Packaging**—People are more and more concerned about food packaging, principally that it does not leach any harmful chemicals into the food and that the packaging itself produces less waste.
11. **We Love Animals**—More consumers are concerned about animal welfare and searching out products that convey improved animal-care practices.

GEOGRAPHIC REACH

"Local" is the new buzzword, and it likely will be for some time to come. People are increasingly interested in keeping money circulating through their local economies and want to see farmers continue to survive in their local landscapes. There are considerable advantages to both a farmer and a community when the sales stay local. However, a farmer may want to diversify the geography of sales to reduce risk, to seek out higher-volume sales, to tap into better-paying markets, and for a suite of other reasons. Many farmers I meet do a combination of local direct sales and regional wholesaling. Some who live quite far from population centers also do direct sales but travel less frequently, selling via a monthly or quarterly buying club (this works better for frozen or nonperishable products).

It is important to think about your geographic reach and all the strings attached, particularly perishability, food safety, and transport logistics. What may at first appear to be a good market may involve more logistics and more costs than you care to entertain. For example, we were once asked to supply a chain grocery store with pasture-raised, organic eggs in all of their Northern California stores. This would have required us to do the following: buy a refrigerated vehicle, maintain and fuel that vehicle, hire and train a driver, insure that driver, call each store for weekly orders, pack those orders, transport those orders over hundreds of miles, invoice those orders, and purchase a more expensive liability policy.

So . . . another vehicle to care for in our fleet, another employee on payroll, a need for year-round consistent egg production (which would lead to a *whole other* group of issues), and increased administration—all to serve these more distant markets. To top it off, these stores paid wholesale prices, usually 60 to 70% of the price we could earn at closer farmers' markets that required none of these added logistics. Think through all these issues before they cost you in time, money, and headaches later on. That is what this book is here for!

SALES VENUES

There are generally three categories of markets you can sell to: **direct markets**, in which you sell to the actual consumer; **food service/ intermediary wholesale markets**, in which you sell to some entity who then sells the food to the consumer; and **brokers,** in which you sell or consign to an entity who then sells it to another market who then sells it to the consumer (it may change hands even more times than this!). Prices received in those different market channels will vary. This is what a typical pricing structure might look like:

Direct Markets (farmstand, U-pick, CSA, farmers' market, buying club, etc.)—100% of consumer dollar

Food Service/Intermediary Markets (restaurants, caterers, cafeterias, direct to retailer)—60 to 70% of consumer dollar

Brokered Markets (broker, distributor, processor, etc.)—40 to 50% of consumer dollar

Therefore, if you sold a bunch of carrots at the farmers' market for $2.00, you could expect to get around $1.40 a bunch if you sold it to a caterer and $0.90 a bunch if you sold it to a produce broker. However, *price* is just one consideration when thinking about your sales venues. You also have to think about postharvest handling requirements, packaging, transport, insurance, and so on when you consider different venues for your products. While the price received at a farmers' market may be alluring, factor in the cost of a market employee (or your own time away from the farm), stall fees, equipment needed to sell at markets (pop-up tent, tables, scale, etc.), vehicle mileage, liability insurance, certificates and inspections, slow sales days or loss, and so forth.

In contrast, selling to a distributor may result in lower prices received, but that distributor might be willing to pick up directly from your farm, eliminating all of the above expenses related to farmers' markets. Penciled out over the year, that may be worth it to you—and you might be able to do more volume sales and perhaps tap into more distant markets. This can be especially pertinent if you live far from any population centers or grow only one or two crops.

Many successful farmers around the country use a mix of the three sales channels, providing a form of risk insurance, diversification of cash flow, more year-round sales and employment, and often a way to get rid of different grades of product (or cuts of meat). Wholesaling is also most common if you are growing nonfood agricultural products, such as medicinal herbs, fiber crops or fiber animals, and others, usually because they require a level of processing that most individual farmers don't have the infrastructure to handle.

RAW OR PROCESSED?

Is there a market for your food in its raw, unadulterated form or in some further processed state? Here's an example: You live near a ski town that also has a short summer tourist season. There is a small summer farmers' market on the town square that you look into selling at. After talking with several vendors, you find out that they sell mostly foods that can be eaten right away or nonperishables that can be taken home on an airplane. That would mean that vegetables that need to be cooked or perishable meats, eggs, and dairy products won't sell well at that market. However, products such as berries, cherry tomatoes, jerky and salami sticks, hard cheeses, breads, and jams and preserves would do very well. Could you diversify into some of these products? Would it pencil out for your business? Would the addition of some processed or nonperishable foods even out your cash flow over the year and provide you with a competitive edge?

If you are considering doing processed foods, begin the lengthy exploration of local and state regulations concerning them. It may take a few months and lots of phone calls before you get a clear answer. Look to your local Cooperative Extension office or Small Business Development Center for advice and support—that is what they are there for. Chapter 13 will go into more detail on value-added products.

There might be growing demand for processed, value-added agricultural products you never thought of before. Here are some that I think have potential:

- Medicinal herb products (tinctures, salves, tea blends, essential oils)
- Dried culinary herbs and herb blends
- Wood products (bamboo, orchard tree firewood, fine hardwood lumber from old orchards, living erosion stakes such as willow, etc.)
- Sugar products (maple syrup, sorghum syrup, natural cane sugar, natural beet sugar, or candies made from any of the above)
- Kettle-cooked chips made from farm-raised root crops (potato, yacon, turnip, sweet potato, yam, yucca, Jerusalem artichoke, etc.)
- Farm-produced fibers (sheep wool, alpaca wool, angora goat wool, mohair sheep, etc.)
- Native plants and seeds for restoration projects, including living erosion mats or straw baffles with native seed embedded in them
- Sun-dried fruits and vegetables (tomato, onion, zucchini "chips," apricots, plums, etc.)
- Anything fermented (sauerkraut, kimchi, pickles, fruit vinegars, cider, and mead)

PRODUCTION ATTRIBUTES AND VALUES

The way you produce your food will help solidify your market niche but, more importantly, will be a living demonstration of your values for the earth, your community, your family, and your own self. You can choose any combination of the following production practices that suit your needs and goals, but remember that they also are a reflection of your values. If you don't agree with a practice based on your values, you should probably set about finding alternatives that better coincide with them; otherwise, you won't have as much pride in what you are doing, and that will ultimately affect your enthusiasm for that enterprise. I have met many farmers who are embarrassed about a certain practice because it is not in line with their values. Yet why produce something that is anything less than excellent? Why degrade your values to sell more of an inferior product? Considering how full of inferior products our modern-day grocery stores are, is there a need to add yet another?

Identify the most important production attributes for you and your customers, then decide if you want to pursue any third-party certifications for those practices. It is by no means necessary to pay an outside party to certify your values, but you may come to realize the importance of those certifications as your marketing and sales expand beyond your immediate circle of friends, family, and neighbors. The further you get from your consumer, the more likely those certifications will become necessary. For example, when we sold our eggs just at farmers' markets, we did not get them certified organic, even though we used all organic practices. In our face-to-face conversations with customers we could explain all of our practices and show them pictures. We also could invite them out to one of our biannual farm tours.

However, when we started wholesaling to grocery stores and no longer had direct relationships with the consumers of our products, we elected to get our eggs certified organic. It was also a practical requirement because we were not allowed to claim, "Fed a 100% certified organic diet" on the egg cartons, so consumers would never know about that essential practice (and a reason our eggs were more expensive).

Likewise, it may be hard to explain all the practices you do, especially with an American populace that in general is ignorant of agriculture. A third-party certification might make that explanation easier for you. You could say that you have implemented all of these practices to improve the lives of your animals, or you can simply say you are "Animal Welfare Approved." You can say that you are reducing your chemical inputs, reducing soil erosion, conserving water, protecting wildlife habitat, paying your employees fair wages, or you could say you are "Food Alliance Certified."

However, before you invest hundreds of dollars (and potentially days of time in paperwork) in a third-party certification program, do some market research on it first. If none of your potential customers can identify that certification, it may not be effective in communicating specific production practices or values. For example, the wine-drinking community has come to embrace certified biodynamic wines with gusto (the amazing quality is probably the main reason), but biodynamic is not as well known among fruit and vegetable consumers. In the following section I will discuss general production values, while delving more deeply into ecocertifications next.

"Organic" is a term that means that chemical fertilizers, pesticides, genetically modified organisms (GMOs), and sewage sludge are excluded from production; synthetic hormones and antibiotics are not allowed for animals; and other soil, water, and wildlife quality–enhancing practices are employed. When annual farm sales of organic products exceed $5,000, certification with an approved certifying agency is required.

"Hormone Free" or "Antibiotic Free" is a term used in animal production. It generally means that no synthetic hormones are used (currently allowed only for beef and dairy cattle) and that subtherapeutic or nontherapeutic antibiotics are not employed. This does not mean that antibiotics can't be used to eliminate a specific disease or pathogen but rather that low dosages of antibiotics are not used in the feed or water of the animal on a regular basis. Currently, there is no government certification of the claim "antibiotic free" except that which is required in certified organic production. This indicates that the potential for fraud is likely with this claim and may create distrust amongst certain consumers.

"Natural" is a marketing term and has no real meaning. For some producers it means hormone and antibiotic free; for others it means minimal processing or additives in processed foods. It has no certification nor any federally approved definition or standard. I suggest not using the term, because it is so ambiguous.

"Grassfed" or "Pasture-Raised" has several meanings, but generally it means that the animal spends its life freely consuming growing vegetation. For 100% grassfed ruminant species (cattle, sheep, goats), the animal needs to consume all of its diet from mother's milk and forages, not grains. This can include hay, baylage, and silage during certain times of the year when grass is not growing. The term "pasture-raised" usually means that the animal lives on pasture and can freely choose to eat pasture plants, but it does not necessarily obtain all of its diet from that pasture. This is especially true for nonruminant species such as pigs and poultry that at best can obtain around 30% of their diets from the fibrous plant materials and so will need additional feed, including protein. There is a newly approved 100% grassfed certification, but there is not one yet for pasture-raised. Some producers are now using the term "forest fed" for pigs raised in forests and woodlands, usually with nut-bearing trees, or "browse fed" for goats raised in brushy areas and eating the kind of mixed diet they prefer.

"Humane Animal Care" means that the animals are cared for under safe, humane conditions that allow the animals to express their natural behaviors, to avoid unnecessary or painful body modifications, and usually to be killed in the quickest, most painless ways. There are a variety of certifiers, including "Certified Humane" and "Animal Welfare Approved." I prefer the AWA certification because it requires outdoor access and restricts confinement-style living conditions, as well as most unnecessary body modifications such as tail-docking of pigs or debeaking of chicks.

"Fair Trade" is a well-established certification for many imported foods (coffee, chocolate, etc.), but it is still rarely used for domestic production. However, there are a number of organizations that will certify a farm as Fair Trade, which usually means that workers are paid fairly and have the right to organize and that benefits accrue equitably up and down the supply chain. Fair Trade, like Union Made, appeals to consumers looking for food products that do not exploit workers and that promote economic justice.

"Union Made" is a fairly small segment of US food production, but it appeals to consumers interested in social justice, employee rights, and economic equality. It means that an approved contract has been signed between the business owners and a union representing the workers. The workers have the right to organize, and the union negotiates on behalf of the employees with the employer. Union contracts are more common with larger farms in which direct communication between employees and employer may be more difficult.

"Carbon Neutral" means that no extra carbon dioxide was emitted into the atmosphere in the production of a foodstuff. For example, a grain producer may fallow a few parcels or add a considerable amount of organic matter in the form of compost to cancel out any carbon they may have emitted during tillage, fuel use, and so on in the production and harvest of that grain. Usually only direct forms of emissions are counted, not the indirect or historic emissions, such as the emissions formed in the production of the compost the farmer purchased, or emissions formed when the tractor was manufactured 20 years ago.

This is a scientifically uncertain formula and not generally used in agriculture except for processed foods. The processing of the food is fairly easy to quantify in terms of emissions, and efforts can be made to eliminate those emissions using renewable energy technologies, alternative fuels, or, in some cases, renewable energy credits (RECs) are purchased to offset the energy and fuel consumption. It might be more honest and reliable for a farm to advertise using "100% renewable energy" or "powered by alternative fuels" than to call themselves "carbon neutral."

"Family Farmed" is another loose term generally used as a marketing tool. A family farm can be a giant corporation that is still controlled by a specific family, or it can be a small sole proprietorship of one or two family members. If you want to demonstrate that your family members are involved member-owners of a farm rather than distant corporate shareholders, you may need to explain this with more than this term. For example, "Family *owned* and *operated* since 1859" might provide a bit more explanation without too many extra words.

"Heritage" or "Heirloom" is also an imprecise term, but it usually refers to growing or raising an older variety of a plant or animal species that is probably not a hybrid, is not genetically engineered, and had qualities that people once held important but that the long-distance supermarket food chain of today might not value. It can be economically challenging to raise some of these varieties or breeds, but they may also provide you with a competitive edge. You may also find that the qualities of these varieties are superior when it comes to flavor, natural disease resistance, climate hardiness, response to daylength, and so on.

ECO/SUSTAINABILITY CERTIFICATIONS

There is a heavy dose of "greenwashing" (a variant of brainwashing) and now "local-washing" going on in Big Ag and Big Food, so how do you convey the honest-to-goodness practices that you use? If you sell direct to the consumer, you could simply talk to them, explain your practices, perhaps show off a few pictures, or invite them out to the farm sometime. As my friends Joe and Julie Morris of Morris Grassfed Beef put it, you are "first-person certified," meaning you don't need any third-party entities certifying the validity of your practices. You let your customers' eyes, ears, and mouths do that. However, as your sales chain expands, or if you start to sell products farther and farther away from your farm, you may want to consider some of the certification programs that demonstrate your commitment to a range of sustainability practices. We touched on these concepts earlier in this chapter; now let's delve into the specifics.

According to Ecolabelindex.com, there are now over 424 ecolabels worldwide. Understandably, consumers are starting to get confused. Here are a few ecolabels that apply to food, that have some consumer recognition in the United States, and that you might consider for your operation:

USDA National Organic Program (NOP)—Federally approved standards of production that prohibit the use of synthetic pesticides, herbicides, fertilizers, antibiotics, and genetically modified organisms (GMOs). You must use a USDA-accredited organization to get certified, which may be available in your region.

Certified Naturally Grown (CNG)—A peer-reviewed, transparent certification that includes all of the "organically grown" provisions in the USDA NOP, along with more stringent requirements that livestock and poultry have access to pasture. Seems to have more participation and recognition in the East and Southeast.

Salmon-Safe—Focused on the West Coast regions that support salmonids, this program certifies vineyards, farms, nurseries, and even urban developments that protect water quality and fish habitat.

The Food Alliance—This certification requires implementation of best practices and continuous improvement in the areas of environment, labor, and animal welfare.

American Grassfed—Requires a diet of only mother's milk and forage for ruminant species for their entire life. Prohibits grain and confinement feeding.

Animal Welfare Approved—Stringent humane animal-care certification; requires outdoor access at all times. Only for family farms.

Carbonfree—A program for a wide range of consumer goods, including food, certifying carbon neutrality. Requires that a company determine its carbon footprint, work to reduce it, and offset the remainder.

The main advantage of ecolabels is that the sponsoring organizations certify or audit the claims that you are making about your products. This can help you differentiate your products from others in the marketplace, perhaps access new markets, retain or expand your market share, and improve your public relations. Read the fine print of the standards before you jump into any of these programs. They can be costly and time-consuming: You don't want to waste your time if you won't be able to meet their standards, or you may even disagree with some of them. For example, the AWA standards for laying hens prohibit the use of hybrid laying breeds. As egg producers we found a couple of hybrid breeds that we liked and that suited our farming system well, so this was not a certification we chose to implement.

Likewise, when we contracted with a farmer to produce organic replacement pullets for us, that producer tipped the beaks ever so slightly to prevent pecking injuries during the brooding phase. Tipped beaks are not allowed in the AWA program, but they are allowed under the USDA National Organic Program. All of these nuances are things you should consider when choosing the specific certification(s) you will pursue.

The following story is of a competent market farmer carving out a space for herself on the East End of Long Island, New York. Marilee Foster is not doing anything fancy or newfangled to market and sell her produce, yet she has developed a recipe that works for her and fits with the culture and consumer dynamics of her region.

Saving Farmland One Row at a Time

MARILEE FOSTER'S FARM, SAGAPONACK, NEW YORK

• Marilee Foster •

Marilee Foster must be wealthy, with all that amazing land her family sits on. Hundreds of acres of prime farmland passed down through at least 10, maybe 12 generations of her family, starting with an old whaling captain who made the wise decision in the mid-1700s to trade his harpoon for a hoe and get himself some land. Although the land had been divided a few times among siblings, cousins, and others, Marilee's dad has pretty much pieced the original Foster homestead back together. Now the wealth is purely the appraised value of the land itself and the amazing crops that it can grow.

When Marilee's dad was young, growing a large, diverse home garden was the norm, along with raising a few meat animals, laying hens, and a cash crop or two. Over the years the cash crop that came to dominate most of the farms on the East End of Long Island was potatoes. They liked the deep, well-draining soils and the climate, and they were easy to ship around the country and even the world. The combined forces of plummeting farmgate prices of the 1970s and '80s and the pressure of developing second homes for the city folk of New York City and its expansive suburbs caused a considerable amount of this prime farmland to get paved over. Topsoil as deep as 16 feet in places now sits beneath behemoth 20,000-square-foot mansions and acres of short-clipped lawn that feeds nobody except the riding lawnmowers.

Marilee grew up working on the family farm, along with her older siblings, Dean and Robin. But later on, her parents didn't exactly encourage her to follow in their footsteps, not wanting her to become a slave to the farm. She went off to a small liberal arts college to study creative writing and art, which she says provides a balance to her life now as a farmer, sort of a yin to the yang. Marilee published a book titled *Dirt Under My Nails* when she was just getting started raising vegetables and now writes a weekly farming- and nature-focused column for her town's local newspaper and a bimonthly one for *Edible East End*, a foodie rag. She also keeps her art going by painting cute little farm signs that customers just adore and throwing a few clay pots in the wintertime when farm life is less busy. She struggles to find balance but finds little ways to maintain a life other than farming.

Marilee came back after college to help her brother produce potatoes and slowly started "dabbling" in raising vegetables, using the odd corners of fields or smaller parcels that Dean didn't need for potatoes. Even though she has been growing vegetables now for 15 years, she said she didn't take it seriously until about 5 years into it. She still refers to it as her "vegetable project" even though it is very clearly a thriving business unto itself. Marilee now rents land from a couple of neighbors who want to see their land in agriculture and from a nearby land trust. The land trust has been a little challenging to work with, since their leases are year-to-year, which doesn't provide her with the security she needs to install a deer fence. With no predators on the island, deer are as prevalent as squirrels. However, she is happy to see as much farmland preserved as possible on the island, which the land trusts have helped make possible.

Nowadays, Marilee grows a wide variety of vegetables, raspberries, and some cut flowers and raises laying hens for eggs. She uses mostly organic practices, although she is not certified. One of her land rentals requires that she use all organic practices, but she has chosen not to get certified to allow her some flexibility in her production. "If a doctor told you that you had swine flu, would you go home and drink herbal tea or would you pull out the big guns to eradicate it?" She has faced the same difficult decision with her tomatoes, which are undoubtedly her most important cash crop. Customers will line up early once her 'maters hit the shelf of her farmstand.

The increasingly erratic weather has created the perfect conditions for the tomato disease late blight, and despite her best prevention methods (five-year crop rotation, pruning, staking, wide spacing, no overhead irrigation, etc.), she has not been satisfied with the only widely accepted organic solution, which is copper. Nonetheless, she doesn't use chemical fertilizers and only on rare occasion does she use a nonorganic pesticide to prevent total crop failure. Still, she stresses that the most important part of pesticide use, organics included, is understanding how it works. The extreme weather variations have her putting more and more of her tomatoes inside hoophouses, where they are better protected from the elements and disease.

Marilee has had a little farmstand on the road in front of her childhood home for over a decade now, selling her veggies, berries, flowers, and eggs. It has turned out to be the mainstay of her business. I asked her how she could compete with the dozens and dozens of larger farmstands around the East End, many of which offer products out of season or from far away. She remarked that her commitment to quality—harvesting fresh each day, selling only what she grows, and growing tasty and unique varieties—has kept her customers loyal. Also, more and more folks are staying on the Island year-round, which allows her stand to be open until nearly Christmas and as early as she can get it open in the spring. Marilee also sells at one nearby farmers' market and wholesales a little to restaurants, caterers, and high-end grocery stores. This year, however, she sold almost all her excess to community farmers' markets, which are staffed with students and volunteers, that local governments have been facilitating to overcome the problem of suburban food deserts. Word-of-mouth marketing is about all she uses—she has no plans to build a website, Facebook fan page, or anything of the like. She doesn't even have a name for her business. She's just Marilee Foster of the Foster Potato Farm. Reputation, good location, and excellent, diverse produce have been the recipe for her success.

MARILEE FOSTER

Who's your farmer?

Marilee hopes to keep farming at essentially the same scale she is at right now. Instead of getting bigger, she is trying to farm smarter. An example of this is that, by hiring an experienced tractor driver this year, she has been able to stay on top of succession planting and weeding more thoroughly. She is trying to find some good help who will work on the value-added processing, such as the jamming, pickling, and potato chip making. In 2003 Marilee and her brother pioneered a microchippery, using only their own potatoes to make tasty, kettle-cooked potato chips. They use only the best sunflower oil and a little sea salt or cracked pepper on the chips. She doesn't exactly want to be peddling snack foods, but the price received for one 5-ounce bag of potato chips is equal to that paid for 20 pounds of potatoes on the wholesale market, so you can understand why they make chips.

Right now the potato chip venture is in a holding pattern. After a few years of wholesaling to a rapidly expanding market, Marilee decided only to retail them from her farmstand; the chips are made seasonally, beginning when they start harvesting in July and ending usually by Thanksgiving. Eventually, they'd like to devote a larger amount of their potatoes to the chippery, but for now the chips are a great exclusive project that can be purchased only at her stand (another way Marilee deals with the tough competition). To make the business bigger, the chip factory needs infrastructure improvements—basically, a larger factory and more automatic equipment. For now, when winter comes it's nice to take a little break; it's the time to repair equipment, plan next year's crops, and maybe do a little art. Who wants to stand over a hot vat of oil all day making chips?

If their plan works, Marilee and her brother should be able to inherit the family farm. Because appraised land values have skyrocketed well beyond what agriculture could pay for, her parents have decided to sell the development rights to a local land trust from a 100-acre parcel of their land. The sale will facilitate Marilee and Dean to own it, farm it, even build a farmstand or farming museum if they so choose, but it will never be parceled out and developed. Essentially, by selling the development rights, her parents will reduce the appraised value by enough that the estate taxes won't be as significant, and Marilee and Dean will be able to afford to inherit the land. The money earned for the easement will be set aside in trust funds to help pay estate taxes on the rest of the farm.

It's a complicated transaction but an important tool for keeping farmers on the land. That said, unfortunately, a lot of preserved agricultural land on the East End of Long Island (the "Hamptons") is not owned by farmers, and plenty of it sits fallow or has been turned into enormous lawns called "orchards" or "scenic vistas." There are an increasing number of new farmers trying to lease out these parcels and get started. They will likely never own the land they farm, but at least they can gain some land tenure to start.

Marilee's advice for new farmers?

- It's easy to plant the seed; now you gotta take care of it.
- Start small; don't assume you know everything—you have a lot to learn.
- Farming is a process of observation along with a process of manipulation.
- You have to be on your toes and be willing to work hard.
- Learn some business management and people management skills.
- Be a stickler about quality.

PRICING STRATEGIES

As a businessperson (yes, that is what you are now), you have to learn how to price your food. It is a balancing act between supply and demand, varying input costs, scale, and production improvements that happen over time as you become a more skilled producer. When you first start out, you may have so little production that you are tempted to either give it away or charge an arm and a leg, because you have so little volume but large startup costs that need to be covered. It's pretty important to get the pricing right from the beginning. It is much harder to have to raise your price significantly because you started too cheap than it is to lower your price as you make improvements and cut costs over time. I will tell you a little story to illustrate how this may occur:

A couple of egg farmers I met this year consistently sold out of eggs the instant they brought them to their farmstand. They also were facing a dilemma of not having enough money to purchase feed for their birds. After discussing their price, I realized they were losing around $1.00 for every dozen eggs they sold, an obvious contributor to their cash flow crunch. To cover their costs and make a little profit, which was needed to cover their fixed costs of running the farmstand, they would have had to raise their price by over $2.00 a dozen, from $3.50 a dozen to at least $5.50 a dozen. However, they felt that their customers would balk at a price increase that large.

They elected instead to raise their price slowly over the next two years to get to the price that made sense (although in two years their costs of production could be even higher, depending on what feed prices do). So in the meantime they would continue to struggle to buy feed for their birds.

If they had done a more thorough job *from the beginning* in figuring out the price they needed to charge, they would now be in a much better financial position, with enough money to pay their input costs and a little extra to cover other business expenses.

Although it's not scientific, I would like you to try the following exercise to determine a fair price for one of the foods you produce. You should eventually do this for each and every item you sell. Using the **cost-plus pricing equation**:

$$P = VC + FC + L + NP$$

where P is Price, VC is Variable Costs, FC is Fixed Costs, L is Loss, and NP is Net Profits

EACH VARIABLE IN MORE DETAIL

VARIABLE	EXPLANATION
Variable Costs to Produce Item	(Total Costs/Amount Produced)
+ % of Fixed Costs	% of total expenses or % of land area
+ Loss % Costs	Deformities, poor quality, death, etc.–figure out cost per item
= Breakeven Price	Your bare minimum price
+ % Net Profit	Whatever profit margin you see fit, maybe 15–20%
= Retail Price	What your price should average out to

So using carrots as an example:

$0.75/bunch of 6 carrots in Variable Costs
+ $0.25/bunch of 6 carrots in Fixed Costs
+ $0.10/bunch due to a 10% Loss
from root worms, deformities, gophers
= $1.10/bunch Breakeven Price
+ $0.165/bunch Profit Margin (15% rate)
= $1.265/bunch Retail Price

Once you have determined your price based on costs of production and profit goals, now you need to reality-check that price in your market. Will you be able to command *at least* that price in direct markets, intermediary markets, or brokered markets? Can you take a little less than that price in some markets and make up for that loss in other markets? Here is an example of how averaging different price points might work for you over the course of a year:

> Over the year say you sell 100 cases of 25 bunches to a wholesale distributor who is willing to pay you $25 a case (equating to $1.00 a bunch, less than your break-even price). The check you receive from them is $2,500. In addition to the wholesale market, over a 20-week farmers' market season, you sell an additional 1,000 bunches of carrots for $2.00 a bunch, for an income of $2,000. Your gross revenues on carrots for the year then is $4,500, breaking down to $1.286 for each bunch of carrots, more than your desired retail price of $1.265.

Keep in mind that averaging prices may not work for you in terms of cash flow. If you won't arrive at your average price until the end of the season, you may have difficulty paying for your ongoing operational costs, which don't change over the season. Additionally, the price of a food item varies over the season based on supply and demand—so you have to feel the market out pricewise in terms of your own demand and supply. One rule of thumb that I live by and suggest all farmers do, too, is: **If you are always selling out early, your price is too low**! Raise your price to the point at which you sell the exact amount of volume you need to during the given time frame.

For example, if you have 100 dozen eggs to sell at a farmers' market on Saturday and four hours to sell all those eggs, raise your price to the point at which you sell those hundred dozen exactly in four hours, not in the first 30 minutes! If being so entrepreneurial feels uncomfortable for you, hire somebody else to do your sales. There may be times that you want to sell your food for less than it costs you to produce it, especially if food access for the poor is one of your goals. If that is the case, look for opportunities in which you can get more than your base retail price at other markets and try averaging the price over the season. As one pioneering organic farmer in California told me, she sells her produce at a couple of "rich markets" to subsidize the lower prices she offers at "poor markets" where she can sell to "her people." I think that is an excellent strategy that both supports your business viability and improves food security where it is needed.

Just don't let your charitable sales drag your business down—we are all better off when you stay a viable farmer. If food charity is what you want to do, consider becoming a nonprofit 501c3 organization so you can receive grants to underwrite your work.

TAKE-HOME MESSAGES

- Don't grow anything without a defined market or without the "market in mind."
- Understand your market demographics as much as possible (maybe they don't eat kohlrabi or okra; maybe they don't like pork but they do like lamb; maybe there is not the income level to support your high production values; etc.).
- Your market is not static. You will lose customers, and you will need to attract new ones constantly. You will have competition—what are you going to do about that? Play defense, attack, or potentially collaborate in some way?
- Try not to sell all of your products through any one sales venue (the proverbial "putting all your eggs in one basket" strategy). Conversely, don't think you have to sell everywhere and run around distributing a small amount of product to lots of different places. This probably doesn't make any financial sense, either, and it will drive you crazy being on the road all the time.
- Don't get into processed, value-added food products without first doing the market research. For example, there may not be the local demand for yet another jam maker, but people may be looking for cut and washed produce. Likewise, when the economy contracts, research shows that people consume more basic staples and do more home cooking rather than eating out or eating processed foods. Will your value-added product be considered a luxury or a staple?
- Don't attempt to compete just on price but rather on values. Unless you are a giant corporation, you should forget about competing on price alone. You are never going to be the lowest-cost producer, and why would you want to be? This book is about creating sustainable, resilient farms, not corner cutters.
- Make your production values clear, and be honest about them.
- Don't set your prices based on what you see in the store or what you think "the market will bear." Your prices should be based on multiple factors, including a real understanding of your

break-even price based on your *real* costs of production. Use the pricing equation I shared to determine what the ideal retail price would look like for each of your products, and regularly revisit this equation as your variable and fixed costs change over time. Don't be afraid to regularly update your prices.

- Do not ask, "How much will you give me?" Instead, state your price and your payment terms, being clear on how many days you can extend credit to a customer. Along with each market venue comes term expectations. The wholesalers and brokers want 30 days; retail usually varies between cash on delivery (COD) and 21 days. Restaurants can often be extremely slow to pay—make your payment needs clear up front.

CHAPTER 3

FINDING AND SECURING LAND

The overall picture of farmland in this nation is pretty bleak: We are losing a million acres a year to development; land prices are going up in nearly every region; ownership is consolidated and more often in the hands of nonfarmers; property taxes and estate taxes can be staggeringly high; and more and more farmers are renting, with no real opportunities in sight for secure land tenure. I have been interested in land tenure issues ever since I read the literature and saw firsthand how lack of land tenure leads to poor resource management decisions. Soil erosion rates are higher on rented ground than on owned land (Fraser, 2004). Likewise, the business viability of being a tenant farmer is often short-lived, with limited options to build equity or retirement security.

On a more positive note, there are more organizations today that help farmers find and secure land and more land trusts conserving farmland. Even land in urban areas is increasingly being converted to gardens and small farms, sometimes on ground that used to be paved over. Incrementally, many of these organizations are seeing their role as facilitating the next generation of farmers and forming a critical link in a sustainable food supply chain, joining a discourse that much of America has entered around local food (Beckett, 2011).

If you are already farming a piece of land that you like, some of this information below will be unnecessary, but you should still read the sections on Family, Owning versus Renting, and Tax and Legal Issues.

FINDING LAND

Although you may drive around and see a considerable amount of land sitting fallow, planted to lawns, or growing low-value crops, it can actually be extremely hard to find suitable farmland to rent or purchase. Here is a laundry list of places to look and people to talk to when searching for land, starting at the microlevel.

1. **Start in your own backyard, front yard, or rooftop!** Start by experimenting in small spaces in and around your house (sprouts on the kitchen counter, potted tomatoes on the flat rooftop!). For example, our first year we purchased about a dozen different breeds of laying hens to raise in our side yard to find out which ones were the most productive egg layers and whether or not they were broody or flighty and to learn other characteristics of their breed. That experimental year next to our house allowed us to settle on a breed that suited our laying-hen operation for years to come once we scaled up to more land.
2. **Try your neighbor, friend, or nearby family member with the 3-acre yard.** See if they would be willing to give up some lawn (and save themselves the time and expense of mowing it all the time!) to allow you to plant a large garden, sow a grain, or even raise a couple of beef steers. You never know until you ask. And they may even be willing to let you experiment on their land rent free (since you know them and all). Even though this all sounds very informal, still write up a little contract stating what piece of ground, how long (end date), what you will grow or raise, and whether or not any form of payment will exchange hands, as well as a dispute-mediation process. Maybe they just want you to mow the other 2 acres of lawn in exchange for your tearing up the third acre. Labor is a great form of exchange, especially when you are cash poor but ability rich! However, you should specify what the labor exchange would look like so the landowner's expectations don't spiral out of control.
3. **Talk to a neighboring farmer.** Perhaps you could rotate a crop or animal with her crops or animals or rent some extra land that she isn't using. One example of this was when we were invited to rent some land of an organic strawberry farmer who had to fallow his fields for five years between strawberry plantings to control soilborne diseases. He would rather not have to pay rent on that land for the five years while he wasn't producing a crop off it,

and he thought the weed control and fertility enhancements of running animals on the fallow ground would be a good idea.

The only reason we did not take the farmer up on this creative idea was that the land was just too far away from where we lived and kept the rest of our animals. The advantage of renting land from an actual farmer is that he will usually have more understanding of what farming looks like and the risks involved. Maybe he will tolerate your junk pile a little better or charge a more equitable rental price than a nonfarming landlord.

4. **Start making a wider circle.** Contact your state land-linking program (www.farmtransition.org) if there is one, and call your local USDA Farm Services Agency office, all your local land trusts, and even state parks. Let all of the local farm organizations know that you are looking—the Grange, the Farm Bureau, the Cattleman's Association, organic certifiers, and any other groups that may be out there. Also, tell every farmer that you know that you are looking for land—much land is rented or sold via word of mouth. I learned one idea from beginning farmer Nathan Winters (see his narrative in chapter 1): He posted a wanted ad on Craigslist that said he was looking for land to rent. His future landlord was looking for a couple who were into gardening and rural living to rent the house—Nathan's wanted ad description fit the bill!
5. **Consider public lands.** One word of caution: although much federal, state, and privately held lands are indeed farmable or appropriate for livestock production, the agencies and organizations controlling that land may struggle to visualize agricultural production on those lands. Perhaps they see farming as a threat to nature or wildlife conservation. It might be necessary for you to educate them on how farming and nature can coexist in relative harmony. Furthermore, if you embrace some form of public access or benefit, such as farm tours and children's education, your farm might be a good fit with the mission of these public land partners.
6. **Work with a developer to find land.** Farming or ranching might also be a good fit in a housing development in which they are required to maintain an element of open space or farmland uses. I have seen models of this in which the houses are clustered up against the trees and the open land in between the houses continues to be farmed under a long-term lease arrangement. Usually, the developer or neighborhood association will determine what kind of production practices may be utilized; however, lower rent prices and longer-term leases often offset these constraints.

GEOGRAPHY AND OTHER LAND CONSIDERATIONS

Where you farm will have a big impact on your economic viability, your choice of crops and animals, your production practices, and your quality of life. It may be convenient to just start farming where you already are, but it might not make sense in the long run. Here are some key farmland variables to consider when making a rental or purchase decision, broken out into farmland production considerations and those beyond the farm fields.

Production Considerations

- **Soil type and quality.** Is it pure clay, clay-loam, loam, sandy-loam, sand, or other? Are there heavy metals, disease pathogens, or other deleterious things in the soil?
- **Acreage available.** Will there be the acreage available at this property or nearby to satisfy your land needs 10 years out? If you want to start small, can you just rent or buy a portion of the land, then expand into more acreage if you decide to scale up?
- **Climate and microclimate.** Is it Mediterranean, temperate, hot and dry, or four real seasons? Is it a frost pocket, a windswept ridgetop, or a cold and wet north-facing slope? Can you practice rain-fed agriculture, or will you need seasonal irrigation? Is there a history of drought?
- **Drainage.** Does it flood regularly? How quickly does it drain? Does it drain too fast? Are there wet areas such as standing water or springs in the fields?
- **Topography.** Is it flat, concave, gentle hills, steep hills, or a mixture?
- **Water source and quality.** Is the source a well, a spring, a creek, a pond, an irrigation ditch, or a city connection, and what is the quality? Are there any bacterial or heavy metal issues? Is there a recent water test you can look at?
- **Irrigation infrastructure.** Is there underground piping with risers, hose bibs, flood gates, well pump, storage tanks, pump from creek or pond?
- **Water costs.** Is the well metered? Will you have to pay for city water? Do you pay for each acre-foot of water from the ditch? What does the electricity cost for pumping? Who controls the

water? (This issue can seriously come back to bite you if you don't do the research: I have some friends that nearly bankrupted their farm when an extreme drought forced them to use expensive city water to irrigate more than usual to keep some tender crops alive. Under normal summer conditions they used about half as much irrigation water.)

Beyond Production Considerations

- **Neighboring land uses.** Is it agricultural? Organic or not? Residential, industrial, wildlife habitat? Will the neighboring uses harm your farm because of thievery, vandalism, pollution, complaints, pesticide drift, and so on?
- **Distance to markets.** I like to draw a two-hour radius around my farm because that is the maximum I am willing to drive to get to markets. Most of my markets will be much closer than that, but an occasional two-hour drive for a good sale might be worth it here and there. On the flip side too close to population centers might limit your land options, too.
- **Real estate prices and development pressure.** Are real estate prices in your area driven up by development pressures, land speculation, high commodity prices, or high-value crops such as wine grapes? How are farmland rental prices or purchase prices behaving? Erratic and inflationary real estate markets might jeopardize your lease or cause your rent prices to rise.
- **Tax rates.** What is the property tax rate for the land? Is the land enrolled in some sort of agricultural value program or long-term protection program to reduce the property tax rates? How much will the property taxes add to the cost of the land? It helps you to know the base rate the landlord needs, because often the renter pays the taxes directly.
- **Quality of life considerations.** These are entirely up to you, but are there people and places nearby that enrich your life? Are there the cultural, educational, recreational, and spiritual opportunities that you desire to balance out your life?
- **Other land types.** Will you have access to or use of a woodlot, sugarbush, rangeland, hunting areas, swimming hole, fishing pond, wildife habitat, and so forth that will bring diversity to your farming enterprise, allow you to be more self-reliant, or improve your quality of life?

FAMILY LAND

Do you have access to family land? Are you already farming on a multigenerational family farm? You are certainly lucky in a lot of respects, especially if you don't have to pay the purchase price of the land. You may still be responsible for paying property taxes, upkeep costs, maybe even the remaining mortgage payments, which might all add up to a considerable financial burden. As some say, you are "land rich and cash poor," but having land tenure is a substantial advantage. On the flip side being part of a multigenerational farm may carry a lot of historical baggage and family drama and might even limit your farming dreams and options.

My husband and I sometimes feel fortunate that we don't have any family members expecting us to farm in the same way they did, to use the same equipment, grow the same crops, utilize the same production practices, market to the same audience, and so forth. We are free to cook up our own farming dream. There are obviously pros and cons to being either a first-generation farmer or a multigenerational one. Let's see if we can maximize the value you may have being part of a family-farming situation.

Land-linking expert Steven Schwartz pointed out to me that family dynamics don't have to be the same as the business dynamics. You can create a culture where there is a time and place for business discussions, setting up defined agendas beforehand. Involve all the family stakeholders in business decisions, including both the on-farm heirs and the off-farm heirs. What do you do if the off-farm heirs want you to sell? Get together and talk about goals. You could sell the land's development rights in the form of a conservation easement, providing the payments to the off-farm heirs to buy out their stake. Or you could purchase life insurance policies for the on-farm heirs, with the beneficiaries being the off-farm heirs (although don't give your siblings too much incentive to be rid of you!). You also might want to consider separating the farmland assets from the business income and assets. That way, if your business does better through all your hard work, the nonfarming heirs won't get to take some of that from you.

The key to having a positive family-farming situation is through regular *communication*. Try to schedule recurring family meetings—once a week is a good idea. Set ground rules for meetings, and try to stick with them, such as giving space in each meeting for everyone to get her opinions in, not talking over each other, active listening, majority rule if things come to a vote, and so on. Perhaps it would be useful to pretend that you all are a board of directors and that you run your meetings according to Robert's Rules of Order. If this helps improve the efficiency of your meetings and limit drama, then it might be worth the extra effort.

Rotate facilitators and note takers, which is also a good way to distribute power. Try to leave personal issues out of the meetings, and focus on your farming and land management goals instead. Set a time limit for each meeting, and table items for the following meeting if you just can't seem to get to them. Write down the key decisions and who is responsible for each action item. There are many other best practices for running a good meeting that can be obtained from the myriad of books out there on the subjects of organizational management, facilitation, and conflict resolution.

The second key issue that can make or break a multigenerational farm is the issue of land ownership and transition planning. Work with an estate lawyer or family business consultant to figure out who owns the land, the business brand, and the equipment. You may want to set up a corporate structure for the ownership of some or all of those things (you can separate the ownership from the decision making and control, depending on how you write up the operating agreement). Make it clear how much equity each family member has in those things, when and how he can sell his equity should he so choose, and how that equity ownership will be passed on in the event of retirement or death.

The worst possible scenario is that a lawyer is charged with dividing up the farm estate because a will and transition plan was never written. This often leads to subdivision and conversion of farmland to development. Along these same lines, estate taxes can be especially burdensome for farmland that has risen dramatically in value (this happens in high-development-pressure regions or when significant farming infrastructure such as barns, greenhouses, and cold storage has been built). Tax laws change each year, but the latest estate tax rate is for holdings worth over $3.5 million dollars; however, there are special provisions for limiting the valuation of farmland to only its agricultural value and not simply using its appraised value. Again, talk to a tax accountant or visit www.irs.gov for more information on how estate taxes might impact your family farm.

OWNING VERSUS RENTING

Part of the allure of farming is the independence—the thought of working for yourself and creating your own wealth through your partnership with the land. Usually, an independent farmer is thought of as a landowner, too, akin to the "Jeffersonian Ideal," but that is increasingly not the case. To have tenure on the land—the ability to hold the land—does not require ownership of the title. There are a variety of ways to hold a piece of land and an increasing amount of creative tenure models, as new and old farmers alike struggle to build viable farming operations. Some of your options include:

Leasing

Leasing a piece of farmland usually occurs in one of two forms: a **cash lease** or a **cropshare lease.** Cropshare leases are more common among farmers who grow commodity crops, while cash leases are more frequent in produce and animal production. A cash lease is a set price per acre, regardless of your production, and is usually payable monthly, quarterly, or once a year. A cropshare lease is a percentage of production, either payable in the commodity itself or in a percentage of the sales of that commodity.

For example, a cropshare lease might require that you give the landlord (who is also a farmer himself) 10% of your wheat harvest or 10% of the income you receive after selling the wheat. A cash lease would stipulate that you pay $100 an acre each year regardless of your wheat yields. You may prefer one method to the other.

Cropshare leases can be more flexible when you are just getting started and are a way for your landlord to share in some of the production risk. Cropsharing landowners may receive a tax advantage if they also complete a Schedule F tax form and write off any losses. However, it usually requires that you keep records that you share with the landlord, something you may not feel comfortable doing. The cash lease is simpler, but if you are just getting started or have any sort of crop failures, you are still responsible for the full rental price, and your landlord does not share in any of those risks.

You may be able to negotiate a hybrid lease that combines some of the advantages of the cash lease and the cropshare lease in one. Perhaps you could build in graduated payments, so that you pay less in the first few years of establishment and more in later years, when you are a more skilled farmer. Or maybe you can negotiate a price break into your lease in the event of a natural disaster.

The length of your lease is incredibly important. If you are just starting out or think that you might move somewhere else in a few years, a year-to-year lease might make sense for you. However, it can be risky, and if land rental prices are increasing, you may be stuck with much higher rent the next time you go to renew. A longer-term lease will allow you to apply for certain government conservation programs, and it may be required when obtaining a bank loan; in addition, you have the benefit of a predictable rent price for the length of the lease. It might also give you the time you need to improve soil fertility, establish pastures, plant perennial crops, and maybe even build some key infrastructure.

Multiyear leases provide landowners and lessees with an incentive to invest in long-term improvements to the land and maintain soil fertility and conservation practices. If you are going to be renting the same piece of ground for 10 or more years, it might make sense to construct outbuildings and irrigation systems and even establish perennial cropping systems such as orchards. It is always a good idea to see what the landowners' goals are and if they might be willing to invest their own money into any long-term improvements or perennial plantings. If not, might they be willing to pay you back for the value of those improvements after depreciation when your lease is over?

The worst place to be in is to invest what little capital you have into infrastructure on rented ground and not be able to recoup any of that investment from the person who owns the land. You might be able to take some of those improvements with you, but that could be more effort than it is worth. You can't really rip up irrigation pipe buried 4 feet into the soil, but you can certainly take the valves off the risers if need be. It would be hard to tear down a barn, but you could probably take smaller skid-mounted shelters with you. Keep those things in mind as you build your business and the key infrastructure needed to support it.

In our own farming situation, we could not build any permanent infrastructure according to the terms of our lease; also, our rented land was in an active floodplain. Therefore, everything we built was mobile by design. We even built a mobile "barn" using a retired cotton trailer as the frame. We could store lots of tools, feed, and other supplies in this structure, lock the door for security, and move it off the property when our lease was up.

Another issue you should think about when crafting a lease agreement is what enterprises you think you might add over the term of the lease. It might be a good idea to get all of those enterprises listed in the lease agreement, along with some words about how each enterprise will require specific infrastructure, equipment, growing practices, animals, and that those things will be allowed under the terms of the lease. I have met several farmers who have upset their landlords or had their leases terminated because it was not clear from the beginning what types of enterprises were going to be conducted, the scale of those enterprises, and what the impacts of those enterprises might look like.

For example, if you are going to start a small dairy herd, you may need manure storage areas, compost windrows, ensiled feeds, and other things that may not smell all that great or always look perfectly clean and tidy. If your landlords do not know this up front, they may balk when you try to add a dairy to your farming business. They may try to terminate your lease or even call the county animal control on you. You never know—the point is that your job is to predict and prevent any possible future problems by crafting a thorough lease agreement that both parties can consent to, and a conflict resolution clause, should any problems arise in the future.

Land Ownership

As for land ownership, I think the most important consideration is whether or not the cost to finance the land is affordable for the farming business. If you must rely on one or two outside jobs to make the bulk of your mortgage payment, then the land price is probably not sustainable. You will wear yourself out working an off-farm job just to pay for the farm (one spouse can certainly work off the farm, but probably not both of you). You also have to think about the consequences of that debt on your financial viability. Will it take 15 years to pay off the debt or 30 years? Will you end up paying double the price of the land to create more affordable payments? If the farm income will not be sufficient to meet the financing payments, you may be better off renting.

According to an Iowa State Extension bulletin, farmers with limited resources might be able to reach a more efficient scale of production by renting versus owning (and this is true nationwide, not just in Iowa). Sometimes it might make more sense to own your home base where you store equipment and have key infrastructure such as outbuildings, greenhouses, or grain silos, while leasing additional acreage for the bulk of the production.

From my conversations with farmers around the country, it would seem that paying between $2,000 and $10,000 an acre for farmland is financially sustainable for a wide variety of crops and livestock production systems. Of course, if there is a residence or other quality infrastructure, you may pay more than that, but if you took off the appraised value of those improvements, would the land price still average within that range? Land ownership can provide a hedge against inflation through appreciation in land values over time. But if you can lock in a rental price that is steady or lower than land appreciation rates, you may be better off renting. As land-linking expert Steven Schwartz says, "Leasing can be a pathway to ownership." Leasing helps you build your business with less risk, gives you time to improve your credit score and debt-to-income position, and brings about other benefits that can eventually lead you into a land ownership situation at some point.

How does a new farmer or a young farmer get herself landed without the advantages of family-owned land, inherited land, or cheap land prices? Coastal California has some of the most coveted land for housing and for farming because of its ideal climate and great soils. Some would suggest moving elsewhere to farm and shipping all the food into the numerous population centers that dot the coast, but you can't beat a climate that lends itself to year-round production of an enormous variety of fruits, vegetables, nuts, livestock, and more.

One young farmer who was determined to grow for the coastal communities around her had to get extremely creative to find all the land she needed for her diversified produce operation. Jamie Collins has worked with private landowners, churches, land trusts, and state parks to piece together the acreage she needs to build and sustain a livelihood for herself and her employees. She is proving that it can be done, as long as you have a good dose of tenacity and an unlimited sense of humor.

Landing Herself in Success

SERENDIPITY ORGANIC FARM, CARMEL, CALIFORNIA

• Jamie Collins •

What do you do when you live in a sprawling, enormous city but have a serious green thumb? Well, first you attempt to grow everything you can in your yard, patio, or tiny apartment deck. Jamie Collins did just that, with tomatoes and grapevines sitting in pots soaking up the glorious sunshine of Los Angeles. She was working in social services but increasingly turning to her plants for nourishment of her soul. Jamie looked into plant science and horticultural programs in her region and settled on California Polytechnic State University in San Luis Obispo, known as Cal Poly SLO for short. She was among the first wave of students there craving training in organic agriculture, and after graduating she went up to Carmel Valley to do research and development for a large organic produce company. She also became an organic inspector for California Certified Organic Farmers (CCOF) and rose in the ranks to become the regional service representative for CCOF, overseeing the work of many other inspectors.

Jamie quickly got the urge to try her own hand at organic farming, leasing a few acres at first in Moss Landing, then moving to the mouth of Carmel Valley. She experimented with many different crops that suited the climate, soils, and local markets. About four years into working two and a half other jobs and farming part time, she was able to transition to mostly farming full time. She still continues to take organic inspection assignments as she has time and if the farming cash flow gets tight. Because of this work ethic and by keeping her day jobs, she was largely able to self-finance the startup costs of her farming business.

The first crops Jamie successfully grew and sold were beets and sunflowers. The Whole Foods Market in Monterey immediately latched onto her sunflowers and became a customer of other crops she grew over the years, until they moved to a central distribution model. (The company has since returned to allowing stores to buy from their local farmers.) Other vegetables such as carrots, salad greens, kale, and chard provided her with the crop diversity to start attending a few local farmers' markets. The warmer climate of Carmel Valley allowed Jamie to expand into heirloom tomatoes, which she did with gusto. Trader Joe's loved her tomatoes so much they even private-labeled them and purchased them for many of their stores in the region.

However, Jamie was just riding the last wave of heirloom tomato obscurity, because soon after, much larger organic farms got into the game and the price dropped through the floor. She took out her one and only USDA farm-operating loan to amplify her wholesale tomato and vegetable production, only to see the tomato market essentially crash. It was a tough lesson and almost forced her to quit. She never wants to endure that kind of loan debt again in her farming operation. From there on out she has stuck to self-financing her farming operation strictly through business sales.

The low prices, strict postharvest handling requirements, the loss of her brand identity, and other disappointments eventually convinced Jamie to forgo wholesaling her produce and work instead to augment her direct sales venues. She still works with one small produce distributor who pays the price Jamie asks for. She is fed up with being a *price taker*—now she wants to determine her own prices based on her own costs and demand.

JAMIE COLLINS

Jamie's colorful tomato display at a food festival

Jamie also sells directly to a couple of restaurants that appreciate her quality, her unique varieties, and how close her farm fields are to their kitchens. She is the Carmel Valley's local organic produce farmer. Ironically, she can't seem to get into the best farmers' markets in her county or the ones adjacent to her. Instead, she drives two hours north to sell at farmers' markets in the San Francisco Bay Area. Although it bothers her that she and her food travel so far, she is thankful for the highly educated and appreciative customer base that resides in the Bay Area.

Even after 10 years' farming, Jamie is still the "new kid on the block" and sits patiently on the waiting lists of her local farmers' markets while vegetable producers with more seniority or political "pull" occupy all the spaces. Absurdly, it might take some weird new food creations she is tinkering with, such as kale dog biscuits and fruit cordials, to finally win her a spot in those local markets. But the core of her farming business—healthy, organic produce—will not be allowed at her booth since other farmers with more seniority already provide the same varieties of produce she grows. This predicament can be witnessed all over the country: newer direct-market farmers clamoring to identify a space for themselves in competitive markets. If we could get more than 4% of Americans buying organic and local food on a regular basis, the pie would obviously expand, and more farmers could have a slice.

Jamie has been operating a CSA for about five years now, serving households

JAMIE COLLINS

Developing new and creative value-added products

on the Monterey Peninsula and the Salinas Valley with 40 weeks of seasonal produce deliveries. She has a core group of CSA customers but must work hard every year to fill her subscriptions, as many people like to hop around and try different CSAs each year or decide to go back to shopping at farmers' markets, where they can select what they want. She has also added seasonal U-pick days with her strawberries, raspberries, and tomatoes to augment her sales and bring people out to her farm.

Involving the community and building more awareness of local agriculture is something Jamie is equally passionate about. She has organized farm dinners, farm tours, movie nights, and a new holiday food craft fair to build that community and help out other farmers and craft food artisans in her area. She eventually wants to create some sort of collaborative effort around food but is not sure yet what that might look like. Would it be a food cooperative, a local farmers-only farmers' market, an online food market, or seasonal on-farm dinners? She doesn't know yet, but her creative mind is spinning with ideas of how to build a more resilient, sustainable food system that supports a diversity of family-scale farmers in the region.

Surviving through the slower winter months can be tough as a farmer, but luckily, Jamie has mastered the art of growing delicious greens throughout the winter, enough so she can still attend several farmers' markets through those gloomy months. In the winter she attends four to five markets, and in the summer she goes to up to six. When you grow produce for farmers' markets, you inevitably end up with product that you take home at the end of the day. Being a creative and avid home cook has enabled Jamie to come up with all sorts of interesting recipes for utilizing that excess produce so none of it goes to waste. She now is renting a commercial kitchen space once or twice a week to process her vegetables and berries into salads, soups, roasted veggies, jams, lemonades, cordials, and more.

One of her latest creations was a massaged kale salad, an intensely healthy and delicious way to eat kale that requires no cooking. Her lemonade blends are growing in popularity, especially in the summertime. And she just might churn out some vegetable-packed dog biscuits to appeal to the growing number of dog owners who fret about their dogs' health. The trick to making value-added products for the farmers' market is that all the ingredients must be your own, with the exception of water, salt, vinegar, and sugar, and of course, they must sell. But making small batches of creative new recipes lowers the risk and gives Jamie the chance to test her markets.

One of Jamie's biggest challenges has been accessing land for farming, but her tenacity, networking skills, and friendly demeanor have helped her form many creative land tenure arrangements over the years. Of course, owning land would be preferable to Jamie, but it has not really been an option in a region where farmland sells for $50,000 to $100,000 an acre! Just recently her luck has turned, because a family member was willing to collaborate on the purchase of a small homestead in Aromas, about 30 minutes north of Jamie's other farmland in Carmel Valley. The family member made the down payment, and Jamie is paying the monthly mortgage payments, taxes, and insurance. Eventually, in about 15 years, she will own the title to the property.

It is not enough acreage for all of her production, but the 5 acres of hilly oak woodland has around 2 acres of cleared terraces where Jamie has planted both annual and perennial crops such as tomatoes and blackberries, along with culinary herbs and lemon and avocado trees, and where she keeps a menagerie of animals for self-subsistence and fun. She can also store all her delivery vehicles, farm equipment, boxes, and seed-starting greenhouses next to the home she shares with her long-time partner. Having a home base has been a huge asset for her sanity and the success of her business.

Farther south Jamie rents fields from three different landlords and has worked with several other nontraditional landlords, too. Currently, from the California state parks system, she rents a few acres of rich floodplain soil that lies in the coastal fog zone, which strawberries love so much. Just a bit inland she rents another 6 acres from a private landowner who will eventually build high-end condos on the fertile soils. Luckily, the housing market has crashed for now, giving Jamie some breathing room for at least a few more years, although her lease is only year-to-year. But it's a risk she is willing to take for land that has been given to her for free.

She also rents another few acres farther up the Carmel Valley that is out of the fog zone and thus can support more warm-season vegetables such as tomatoes, peppers, and the like. Again, this lease is year to year, and she pays cash rent for it. As an organic farmer Jamie must first get the land certified organic, which takes considerable paperwork, money, and sometimes a wait of several years if it is transitioning out of conventional production. Then she invests in compost, cover cropping, irrigation systems, trellising, and other improvements to support her organic production, but she does all this with the risks inherent in a one-year lease. It has certainly not been easy.

Her state parks lease is a bit longer, which is needed for perennial strawberries, but the time and complexities it took to negotiate with the state more than offset the benefits of a longer lease. (It took three years for them to finally approve the lease, since they had never dealt with farmers before.) Altogether Jamie cultivates around 16 acres of land on four different properties and provides employment to a crew of four farm workers year-round. She works hard to provide year-round employment, and her workers have remained loyal in part because of that commitment. Jamie also employs several other part-timers, many of them friends, who help her staff the numerous farmers' markets she attends each week.

Not having secure land tenure has prevented Jamie from planting many of the perennial crops she would like to diversify into. In the case of her raspberries, she has had to relocate them three times now, something she said she will "never do again!" Fortunately, her new homestead land gives her the ability to plant some of the cane berry and tree crops

Growing food among the mansions in the Carmel Valley

that she has been dreaming about. It also turns out that avocados and lemons can be the ticket to get into some of the more coveted farmers' markets of the region.

Land zoning classifications have also gotten in the way of some of her dreams. On one property, in which she was negotiating with the landowner (a church) to rent some of its land in exchange for food for the poor and garden education, progress was halted because of the complaining of a neighbor. This neighbor did not want anything to be built or grown adjacent to her home, and her complaints to the county reminded them of their recent rezoning of the land for high-density condos, thus preventing Jamie from moving forward with planting. Unfortunately, she had already invested in some soil preparation and deer fencing when the farming project was shut down.

Jamie also rented farmland for a few years from a local land trust but did not find the support she expected from them. The trust only wanted to rent her a large contiguous block of land—considerably more acreage than she needed or wanted to pay for. The land trust owned half of the land, only a third of the water rights, and no infrastructure; therefore, they were unable to give the support they could have. The access road to the farm ran right next to several small cabins that were rented out by the neighboring landowner. Strangely, even though the people were living in former farm worker cabins on a historic farm, those tenants complained about the dust generated by Jamie's farming activities, harassed her customers who came to pick up on the farm, let their dogs run through her fields, and even stole food out of the fields. Considering she was the poster-child farmer for this land trust and was simply a renter doing what she was allowed to do with their land, it is disconcerting to hear that the land trust did not intervene to protect her from harassment and also wasn't more flexible about their leasing. Sadly, that field is now sitting fallow, filling up with invasive weeds, instead of growing a healthy variety of organic vegetables as it once did under Jamie's management.

Another dream of Jamie's is to have a retail farmstand located on one of her farm parcels in the Carmel Valley. Again, the lack of land tenure has prevented this idea from moving forward; however, she has a new idea that just might solve this problem. With a box truck that she no longer uses for wholesale deliveries, she is considering creating a mobile farmstand that she can move around depending on the season and where the most customer traffic is.

Luckily, Jamie's ideas and voice are being given more space now that she writes for the *Monterey County Weekly* newspaper and the quarterly *Edible Monterey Bay* magazine. Eventually, she would like to write a book about farming and her story, but the time for that will have to come down the road: Right now she is just too busy growing food for people and trying to find and secure the land she needs to do that.

Jamie's advice for new farmers?

- Start small. Develop your markets before you grow bigger. Make sure you know how to produce well before you quit your day job.

- Direct market as much as possible. Don't let the middleman take your money.
- Connect with the community around you.
- Do high-value things, niche things. Use whatever you have, and make them into products. There are a lot of value-added possibilities if you are creative.

ALTERNATIVE TENURE MODELS

Jamie Collins in the narrative above turned to some nontraditional landlords to find the acreage she needed to grow the diversity of vegetables she wanted and to be able to rotate her fields and rest them periodically. There are many nontraditional landlords and land tenure arrangements to be found out there that could give you more options in your farm production. Ideally, you won't be in a rush to get land and can take the time to consider all sorts of creative arrangements.

Some Nontraditional Landlords

- Privately owned or city-owned unused urban lots
- Hospitals with extra land
- Churches with extra land
- Public and private schools with extra land
- Federal lands (Bureau of Land Management, US Forest Service, etc.—this is more common with grazing leases)
- State lands (state parks, watershed lands, etc.)
- County lands (parks, watershed lands, redevelopment lands, etc.)
- City lands (parks, redevelopment lands, etc.)
- Land trust lands
- Utility-owned lands
- Housing developments with common land
- Cohousing and intentional communities

Some Nontraditional Land Tenure Arrangements

- **Food-exchange lease.** You provide food to the landlords in exchange for using their land. A church, for example, might want you to grow food for the poor in their congregation in exchange for using their land. A private landowner might want a CSA share in exchange for using his land.

- **Education-exchange lease.** You provide training and education in exchange for using land. A city park, for example, might allow you to use part of their land in exchange for teaching gardening education courses for their community garden. An elementary school might allow you to use an acre of fallow land at the back of its property in exchange for providing science-based garden education classes to its pupils.
- **Restoration lease.** You restore all or part of a piece of land in exchange for farming part of the land. A land trust, for example, might lease some pasture to you in exchange for your using management-intensive grazing to improve its stands of native, perennial grasses. A city might allow you to farm an abandoned lot in exchange for thoroughly cleaning it up and remediating the soil.
- **Fix-it/maintenance lease.** You maintain or fix certain things on a property in exchange for farming part of the land. Perhaps you caretake a property when the owners are away, maintaining all of their equipment or keeping their plumbing systems from freezing, in exchange for using some of their land to farm. Alternatively, a utility company, for example, might lease you the land under their electric transmission lines to graze your animals. You in turn keep the power lines free of tall vegetation.
- **Rent-to-own.** This option might work with landowners who are considering owner financing or those that really want to see their agricultural business continue. This is similar to owner financing, but perhaps you could arrange lower interest rates than typical owner financing and get rid of the giant balloon payment that is often part of owner financing (e.g., pay $10,000 a year for five years, then pay off the rest in a big balloon payment).
- **Owner-financing or installment sale.** Make a down payment, then pay set monthly payments for several years, and finally pay off the final balloon payment through refinancing. If you miss a payment, the landowner can often foreclose on you. Be cautious about this type of arrangement because you may not be able to get a conventional mortgage in time to pay off the balloon payment, especially if it is just raw land with no house. Banks are not loaning much on raw land these days (agricultural banks, on the other hand, may consider this).
- **Work-to-own.** This arrangement is similar to the one above in which you rent to own, but in this scenario you work for the farmer for many years with low to no salary (perhaps living for free on the farmer's property) in exchange for an increasing ownership stake

in the business, until you eventually own it. I have seen arrangements that are staggered over time, such as your being paid 10% of a manager's salary in year 1, 20% in year 2, 30% in year 3, and so on, until year 10 when you are paid the full salary and become the full owner. These arrangements are more common for the transfer of the business, but not necessarily the land and its improvements.

- **Purchase land with a conservation easement.** Say you find a piece of land that has significant natural habitat or is under imminent threat from development—a piece of land that a local land trust organization might be interested in. Perhaps they could purchase the value of the development rights or the conservation easement value (which is normally the value of the natural habitat) while you raise the rest of the purchase price. The land trust's contribution could be your down payment, maybe allowing you to qualify for a lower-interest-rate mortgage or lower monthly payments. Selling the development rights can also lower your property tax obligations in the future.

  This is a complicated process and requires a patient seller willing to take the time, but it just might be a good fit when you find a conscientious seller who wants to see the land spared from development or remain in agriculture. In addition, a land trust may already own the land and want to sell it to a conservation-minded farmer.
- **You own the improvements; the land trust owns the land.** This is a new ownership arrangement that a few land trusts are employing, particularly in markets with high-cost farmland that few farmers can afford. In this arrangement the farmer purchases the improvements, such as the house and barns, while the land trust owns the land underneath them. This allows the farmer to maintain and improve key infrastructure, live on the farm, and build some equity without the burden of financing all of the farmland, which might prove too costly for him. The land trust then leases back the farmland to the farmer at an affordable price, usually in the form of a 99-year lease. A land trust in Massachusetts called Equity Trust has pioneered this model, and it is one that I hope to see expanded. It takes a special kind of land trust, one that makes it its mission to see farmers on its land and provide opportunity for new generations of farmers, too. Maybe you can talk your local land trust into exploring this idea.
- **Set up a site-specific community land trust** for the farm you want to buy. Live Power Community Farm in Covelo, California,

is an example of this. Their CSA customers formed a community land trust to purchase the farm for the owners.

- **Purchase a small homestead and lease neighboring property.** This situation is quite common but often overlooked. You could purchase a small homestead, say 1 or 2 acres, in which to live, store equipment, breed your animals, build your seed-starting greenhouse, and so on, then lease the bulk of your farming acreage nearby. Perhaps you can lease the neighbor's property so you don't have to drive far. If this idea suits you, you should talk to the neighbors before you buy to make sure that you will actually be able to lease their land under favorable terms. We have met many farmers who were able to negotiate positive long-term leases on neighboring properties and are now mortgage free and debt free. Remember, you don't have to own all the land you need to farm—the key is to be strategic about it.
- **Collaborate with others to purchase land.** This could be in the form of a nonprofit organization, a cooperative, or a corporate-type structure such as a limited liability corporation (LLC). It is essential that you get assistance from a good lawyer when crafting this kind of arrangement and build "exit strategies" into the agreement in case anybody wants to leave, or a divorce (of sorts) happens after some time. But it just might be the ticket to making farmland affordable and gathering together the right skill sets for creating a sustainable, diversified farming and food business. Imagine a group of you each putting in $20,000, each with different skills that complement each other, creating a much more stable business than if all of you attempted to go it alone.
- **Customer-financed land purchase.** I will explain this in more detail in chapter 4 when discussing creative financing, but you might be able to get loans and prepurchase payments from customers that would allow you to purchase a piece of land. In one farm model highlighted in chapter 5 (Misty Brook Farm in Massachusetts), around 15 customers loaned all of the financing needed to purchase a small homestead for the family, liberating them from the hassle of trying to get a conventional bank mortgage for a couple of self-employed farmers. In another potential model the aggregated prepayments from customers could provide the down payment that a farmer needs to qualify for a bank loan.
- **Lease from developers or banks** that have foreclosed properties (real estate owned or REOs). You will need extreme patience however with this process because they often have a backlog of

properties to deal with and leasing farmland may not be a high priority for them. Remind them how farming their land will decrease the weeds, fire hazards, and even prevent vandalism that happens on many empty bank-owned properties.

TAKE-HOME MESSAGES

- Don't rely on just a handshake agreement when renting land. A written land contract is the most important piece of paper you can have when farming, even if it's a simple one-pager. Even if you are just renting a piece of ground to hay it and your "rent" payment is a percentage of the hay bales, write up a one-page contract stating exactly how many bales you will provide, and have the landowner and you sign and date it.
- Aim to have a thorough understanding of the landowner's goals, needs, and wants and also be clear about your own (this helps manage expectations). For example, if the landowners plan to sell the property for development in three years and just want some rental income in the meantime to pay the property taxes, they probably won't value any improvements you make to the soil, fencing, irrigation system, infrastructure, and so on.

  Likewise, the landlords may have a different vision for their farm from the way you want to go about farming. They may be looking more for a landscaper, someone to "tidy up" the place, prune their orchards, repair their rock walls, fix their roads and culverts, thin their woodlot, and the like. You may not be "tidy" enough for them; for instance, they may not like the look of your equipment yard, irrigation pipe, scrap woodpile, or the plastic-covered hoophouse that you hope to build. It is extremely important to get those things out in the open *before* you sign a contract with them. Try to think of every last enterprise and practice that you might try so they understand fully the potential impact of your farming operations.
- Likewise, share with them some of the benefits you can provide, such as improving soil organic matter, repairing major soil deficiencies such as a calcium-to-magnesium imbalance, cleaning up of fence lines, or removing trash in the fields. Even if the person you might rent from is a family member or friend, still have a thorough conversation with her about her goals, needs, vision, and so forth.

- Be clear about dates and terms of the lease and when to renew it. Where can you get some templates for model leases? Try Start2Farm.gov, growingnewfarmers.org, or contact your state land-linking organization to request model lease templates.
- When leasing land, shy away from building infrastructure or making other large investments without an understanding of how you will be paid back or build equity in those things.
- Match your crops or animals appropriately to the land you buy or lease. Consider such things as soils, climate, topography, and irrigation.
- Have a good understanding of water quantity, quality, source, and rights when signing a lease or buying farmland.
- With family-owned land have clear, regular communications about the land ownership and management; likewise, have a transition plan.
- Don't think you have to own *all* of the land you need for production. It might make more sense to own some, especially where your key infrastructure will be located, and rent some, depending on the cost of farmland, property taxes, and so on. Consider some of the unique, creative land tenure situations I've mentioned above to access the land you need.

CHAPTER 4

FINANCING THE DREAM

How do you come up with the money to start a farm, expand production, or start a new enterprise? Financing is one of my favorite topics. Get ready to think outside the cornfield.

There are generally two classes of financing: **working capital** for operational expenses and **long-term capital** for asset expenses. **Assets** are items such as equipment, breeding stock, and infrastructure that will serve your farming business for many years to come (sometimes they are further classified into **midterm** and **long-term assets**). Working capital is usually paid off during the year of production or at the very end of that year, whereas long-term capital is paid off over several years. Startup capital is usually a blend of both working and long-term capital because a beginning farm often needs basic inputs such as seeds, feeds, and fuel, as well as some key pieces of equipment or infrastructure.

The following paragraphs discuss a myriad of ways to obtain both working and long-term capital, their pros and cons, and some specific examples or anecdotes of these forms of financing in action. I should point out that the term *financing* means money that will be paid back. Later on in this chapter I discuss money that does not have to be paid back, in the form of grants and donations. I am constantly asked by farmers about free sources of money: They are few and far between and actually take a considerable amount of work to obtain—not exactly what you would call "free" money.

CONVENTIONAL DEBT FINANCING

With conventional debt financing you receive a loan and pay it back over time, usually with interest payments. Interest rates can be fixed over the life of the loan or variable, meaning they can go up and down. Variable interest rates are more common with mortgage loans but are not often seen in production loans. Loans can be obtained from banks, credit unions, and some governmental agencies, such as the USDA Farm Services Agency (FSA). There are also dedicated agricultural lenders such as Farm Credit and some regional banks that focus on agricultural producers. Nonagricultural banks may have a hard time understanding what you want to do and may be less likely to take the risk on an agricultural loan.

Advantages of debt (loan) financing are the fairly quick turnaround between applying and receiving the loan (if you qualify), the predictable payments over time (unless you elect a variable interest rate, although you may not have a choice), and amortizing payments for capital investments over time (often the useful life of the investment)—and if you pay the loan in a timely manner, it can improve your credit score and help you qualify for future loans.

Disadvantages can include a lengthy application, difficulty convincing a bank that agriculture is a good investment, potential high interest payments, and risk of crop or market failure (making it hard to pay back the loan); also, sometimes the banks try to stipulate what you do with the loan and potentially stifle some of your creativity or thriftiness (such as specifying that you can't build your own barn and have to work with a licensed general contractor instead).

It is hard to qualify for a loan in your startup phase without a proven track record, and most lenders don't want to make loans under $250,000 simply because it takes the same amount of time to put together a million-dollar loan as it does a small loan. (Lenders actually think $250,000 and under is a small loan!). They will also often require that you personally guarantee the loan and put up collateral as your guarantee, but when you are a new farmer or have limited financial resources, you may not have many assets to put up as collateral. Likewise, the one or two assets you do have you may have already promised to another creditor. For example, if you have an automobile loan for your farm truck, you can't put that up as collateral for a different loan. It is already promised to the bank that you have the auto loan with.

It's worth noting that the farmers we have met in our nationwide amblings are pretty much equally divided on the issue of debt financing. Half of them tell us to never take on debt, especially when starting up a farm. The other half tell us to use debt strategically, for investments that will improve your bottom line. However, almost all the farmers we interviewed have used debt financing a couple of times for specific projects over the life of their farm. Many carry a mortgage, but one they can pay for with their farm earnings. Others have taken on midterm loans in the three- to five-year range for strategic capital investments, often to expand their production or add value to their existing production with specialized processing equipment.

One farm was building an improved yogurt-packaging line with a loan. Another had taken out a 36-month loan for a used box truck to deliver their produce and flowers. Still another had a 5-year loan to buy a mobile commercial kitchen to process their meats and vegetables and sell them at various events. With the exception of one farm I profile in this book (Hoch Orchard & Gardens in Minnesota), none of the farms we visited regularly took out operational loans each year, although many did have long-term mortgages to finance their farmland and homes. Most self-financed their yearly operational expenses using the proceeds from the year before. I don't know if this is a trend among forward-thinking, sustainable farmers, but it appears to be a characteristic that holds true for all the farmers I spoke to. It seems to be the larger commodity growers or contract farmers that take out the large, yearly operational loans, but that kind of risk is often avoided by the more diversified, direct-market farmers and those of smaller scale.

However, if you have been in business for many years and have a fairly predictable market and an excellent handle on your finances, particularly cash flow, taking out a modest operating loan each season and paying it back within the same year might be a good idea for you. An operating loan can certainly eliminate much of the early-season cash flow crunch and allow you to move your best production forward. There are other options to conventional debt financing, which I will discuss below.

PERSONAL FINANCING

Personal financing is very common with new businesses that haven't established much of a track record and don't often quality for regular loans. Essentially, you use the savings that you have built working other jobs, or use the ongoing salary that you receive for your off-farm employment, to pay for operational expenses or long-term assets. If you take money out of your savings or retirement account to start a farm or start a new enterprise, it is wise to try to pay back those savings over time, just as you would a loan. Otherwise, you will deplete your savings and not have that money for a rainy day, retirement, a medical emergency, or other challenges that inevitably come about in life. If you use your off-farm income to pay for farming expenses, try not to make that a habit. Endeavor to pay yourself back by having the farming business pay for some of your living expenses, or use that money for strategic longer-term investments that have a decent return.

Some advantages of personal financing are these: Obviously, you don't have to fill out a 30-page application to get money from yourself or your spouse/partner (however, you should still write up a decent business plan); you don't have to pay yourself interest; the risk is lower; and it won't negatively impact your credit score. Some disadvantages of personal financing are that you could quickly deplete your savings or retirement account; you could become reliant on the cash infusion from your off-farm employment and quickly work yourself to the bone holding down multiple jobs; and it could drive a wedge between you and your spouse/partner if he is helping to finance the business or paying all of your collective living expenses. You also don't improve your credit score when you use only personal financing because it is not reported to a credit agency.

There are many stories of farmers using personal financing, and it's probably the most common source of cash to start a new farm or new enterprise. In one scenario the spouse's income can be used to pay for most of the living expenses, such as rent or mortgage, groceries, vehicle, insurance premiums, clothes, and vacations, while the farming profits are turned back into the farming business and don't have to be used for family expenses. This situation works best when there is good communication between spouses/partners and when the spouse working off-farm is content with this arrangement. However, if the spouse working off the farm eventually wants to work *on* the farm, you must communicate and create a plan to increase the farm income such that it can cover household living expenses. Don't ever make assumptions that this arrangement will carry on indefinitely into the future.

I have also met many farmers who built a nest egg over time working in another career and who utilize that money as either a down payment to purchase a farm or utilize those savings for essential startup equipment and materials that will serve them for many years. There is nothing wrong with doing this; I only suggest trying to rebuild that savings back up over time to where it was. Pay yourself back as you would any lender—perhaps even with a little interest.

INVESTOR FINANCING

Investor financing is more common for fast-growing, high-revenue businesses and is not usually a good fit for most family farms. Private investors provide money in the form of stock shares or cash infusions, but unlike with a conventional loan, the investors do not get paid back the original investment plus interest. Instead, they either own a portion of the business or are entitled to a percentage of the business proceeds over the life of that business. This is a complicated form of financing and requires incorporation of the business to receive those investments.

Advantages include the ability to raise large amounts of capital, often larger than can be achieved with traditional debt financing, and the investor only gets paid if the business is profitable. Likewise, private investors are more tied to the business, and consequently, they often become more involved; for instance, by serving on an advisory board or board of directors. Their expertise may prove to be invaluable, or potentially annoying.

Disadvantages of investor financing are the extensive legal filings that must be submitted to the federal government and to the government of any state in which the offering will be made; the long-term commitment and relationship with the investors (this might be particularly detrimental if their values are not aligned with yours); the possibility that investors might force you to give up control or, worse, actually sell the business out from underneath you; and the profits you have to pay the investors might take a huge bite out of any profits you may make. The majority of "mainstream" investors expect really high returns, upward of 20% in as few as five to seven years. It is no accident that most progressive businesses are privately held and not owned by outside investors (shareholders). The normal assumption of publically owned firms is that they will value making profits above all else, usually an unacceptable standard for triple-bottom-line businesses and those based on the natural capacities of the soil.

However, equity investment might work for some farming operations; for example, those that are vertically integrating and that are going to capture significant market share as a result, or those farms that have developed a new technology or value-added product that has high growth potential. That business could be spun off and invite in equity investors. That would also help protect the actual farmland from any forced liquidity event that the investors might demand. Say, for example, that a farmer designs and builds a new grain-milling system that is cheaper to run and creates less waste and better quality. Perhaps other farmers want to buy that milling system, but you need investors to manufacture it on a larger scale. That might be the appropriate enterprise to draw in equity investors to scale up the manufacturing quickly before somebody else patents and markets it.

Another model of bringing in outside investors without losing control is what the Cooperative Regions of Organic Producer Pools (CROPP) Cooperative has done (Organic Valley brand). Starting in 2004, CROPP began selling what is called "Class E, Series 1 preferred stock." This kind of stock is not allowed in some states, so they began with the states that allowed it and have expanded as other states get on board. This nonvoting stock allows supporters of the CROPP Cooperative to receive a fair return on investment (6%) while supporting a company and a mission they believe in. Having the investment without all the strings of outside investor control attached has been a great boon to the cooperative. As long as the farmer-owners continue to maintain the majority of the shares (51%), they remain firmly in control of the business.

Find out from your secretary of state's office what is allowed for your type of business.

CONSUMER AND BUYER FINANCING

Consumer and buyer financing is sort of a hybrid between debt financing and investor financing in which your customers finance you. This can take on many forms such as prepurchases in which the customers pay for something they have yet to receive (such as CSA subscriptions or gift certificates); private loans, in which the customers are paid back in cash, product, or both; or when a customer purchases equipment, breeding stock, or other inputs for your business with an advanced contract. An example of this might be a grocery store that buys a refrigerated truck for a farmer in exchange for locking in a volume and price for eggs until the truck is paid off. Regardless of the arrangement you negotiate with your customer(s), develop a written contract that specifies all of the terms, dates, and so on. There are new Federal Reserve Board regulations governing gift certificates and gift cards—make sure you understand what those rules are and how they apply to your business.

Advantages of consumer financing are that it requires a relatively light load of paperwork (although there are laws governing these forms of financing); it has a fairly fast turnaround; potentially stronger relationships may develop between you and your customers; and interest rates are often lower and terms are more negotiable, so as to be more in line with the cash flow needs of your farming business.

Disadvantages include the risk that you won't be able to pay back investors because of crop failure or other natural disasters (this happens in some CSA farms when a crop failure does not allow them to deliver product to their shareholders); the potential that sticky money issues might drive a wedge between you and your customers; the fact that this kind of financing does not build your official credit score; and the problems that might arise when you still owe money or product to a customer, even though you may not want to do business with him anymore, for whatever reason. Perhaps you found more lucrative markets and no longer want to sell at the wholesale prices you locked into your contract, or you owe your customers a product but you no longer want to have that particular enterprise. However, if you have made a commitment, you have to stick with it. For that reason, it is important to look at the length of the contract and whether or not it will suit your business over the full term of the contract.

A common example of this kind of financing is selling CSA shares, in which the customer pays you up front for a share of your farm's harvest over the season. Some CSAs divide up that payment over the year into quarterly or monthly installments so the cost is not too much to bear for lower-income customers. Another example of customer financing can be found in the Misty Brook Farm narrative in chapter 5; they are financing their homestead with private loans from 15 customers. The owners of Misty Brook are paying some of their investors in cash, some in cash and product, and still others with just farm products. In yet another example, which I will expand upon in the narrative profile section of this chapter, a large CSA farm made a loan to a chicken farm in the winter so the farmer could purchase a new flock of replacement pullets. The loan was paid back over the egg-laying season with a reduced price and set volume of eggs.

Be aware that private loans can be subject to certain securities laws in your state and at the federal level. I don't want to scare you away from private loans, because they are an excellent source of farm financing, and they give people an opportunity to invest their money where their values are. At the very least, though, use a legal promissory note agreement and make sure you abide by the terms. (You can often find generic promissory note agreements on the Web that are approved in your state, or search out Nolo's book on the subject, titled *Business Loans from Family & Friends.*) If you don't pay back your lender, you may be opening yourself up to legal trouble. Additionally, if you are inviting in private lenders, don't advertise it broadly. Stick to your core customer base, and let them know verbally or through a private e-mail list. This makes it easier to defend that your loans were made through a private offering, not a public one, which opens you up to securities law.

Outside your core group of family, friends, and customers, there may be private lenders you can tap into through the Slow Money network. Although this nascent organization has no financial instruments of its own, it has many members who are looking for places to invest in that are in line with their values. One farmer I know in California was able to meet a few private lenders via a Slow Money chapter in the San Francisco Bay Area. Slow Money also hosts a large conference every year or two where they bring together investors and food-based businesses. It may be worth attending to refine your "pitch," as well as to meet potential investors. Remember, though, that if you can get even "slower" money through a standard farm loan, you should probably go that route.

GOVERNMENT GRANTS AND LOANS

There are also a number of government grants, loans, and loan guarantees out there that are worth researching. I won't go into extensive detail about them because they are constantly changing and, by the time this book gets published, my information might be out of date. At the city or county level, there may be economic development incentive programs, loans, or tax incentives, especially if you plan to create jobs. At the state level there are also economic development departments and departments of agriculture that could have loan or grant programs. Many of these programs focus on value-added processing and job creation. Do not be shy about tapping into these resources; they often go to larger corporations, but there is no reason that small farming businesses should not benefit from them as well. You can help divert these monies to building regional sustainability and food security!

At the federal level there are several agencies with funding programs that can directly or indirectly help farmers. Most obvious is the US Department of Agriculture (USDA) Farm Services Agency (FSA). This agency offers shorter-term operating loans (OL) and longer-term farm ownership (FO) loans up to $300,000. They reserve a portion of these funds for beginning and socially disadvantaged producers (essentially, anyone who is not a white male). FSA loan interest rates are quite low, but the paperwork process tends to be onerous and the wait times long. You also typically have to be rejected by several mainstream agricultural lenders before the FSA will consider you, but they also offer loan guarantees (which means you can go to a commercial bank with a guarantee by the agency), so mainstream lenders might take you on as a higher-risk client.

The USDA also has several competitive grant programs that can fund aspects of your business and on-farm research. These include the Value-Added Producer Grants (VAPG), which fund planning and implementation activities related to adding value to your farm products, marketing, and some agritourism activities. Another is called the Rural Energy for America Program (REAP), offering grants and loans for on-farm energy efficiency and renewable energy projects.

An additional federal agency that offers loan guarantees is the Small Business Administration (SBA). They can help fund business real estate purchases (including farms), inventory, equipment, and more. They also provide business technical assistance, such as help with crafting a business plan and finding a business mentor in your community. They may not have expertise with farming per se, but they do in many other aspects of business management that are essential to all farmers, such as planning, marketing, and financial management.

OTHER CREATIVE MODELS

Nowadays there are several "crowd funding" websites available to you, where you raise small amounts of cash from lots of motivated people, often strangers. They are split into sites that gather donations that you don't have to pay back and sites that offer loans that you do have to pay back. Some current donation-based websites, which may already be out of fashion by the time this book gets published, are Kickstarter.com and Indiegogo.com. First, you have to pitch your idea to the organization itself, usually in the form of a video. Putting together a polished or catchy video might not be part of your skill set, but perhaps a family member or eager customer of yours would be willing to lend her skills.

If the organization accepts your pitch, you have a set amount of time to get the word out and convince people to make donations to fund your idea. If your idea is very localized in scale, it may be hard to get people from all over the country to pitch in. You may have to offer compelling gifts in exchange for different donation levels, such as inclusion on your website, a copy of your new cookbook, or a private farm tour. In the case of Kickstarter, if you don't reach your fund-raising goal, you don't get to keep any of the pledged donations. With Indiegogo, you get to keep whatever donations you have raised by the cutoff date.

I find that people are more apt to give to a campaign that has further-reaching impacts than just your little business. For example, say you were going to open a self-serve farmstand that allowed people to pay what they could on a sliding scale so that all people in your community, regardless of income, could have access to fresh produce. That project might be more compelling because of its social benefit than a campaign to build your farm a solar dehydrator so you can make more money in the off-season with dried produce. Just a thought . . . .

There are also websites where you can attempt to crowd-fund your business through small loans. These include Prosper.com and LendingClub.com. Friends, family, and strangers all over the country can lend as little as $25 or into the thousands, while earning some interest on their investment. Prosper.com takes on higher-risk loans but charges very high interest rates, whereas LendingClub is very selective about whom they offer loans to and thus charge lower interest rates. If you can work out small loans among friends and family without having to resort to these for-profit lenders, you will likely pay less interest and administrative fees over time. Likewise, many bank loans and credit cards have interest rates lower than what I have seen on Prosper or LendingClub, and if you pay them on time, they will positively impact your credit rating.

Your community or region may have a revolving loan fund offered through an accredited Community Development Finance Institute

(CDFI) or a Micro-Enterprise Development Organization (MDO). Ask around, particularly at the closest Small Business Development Center, about what might be available in your area.

For this chapter on financing and funding your farming business, I chose to profile our own farm, partly because I am so familiar with the story, and partly because I didn't find any farmers who were willing to divulge this level of financial detail to me. By no means are we unique or that financially savvy, but we were willing to get creative, utilizing some ideas and some variations on others to put together the funding we needed to scale up our business to a respectable midsize farm.

Our Multipronged Financing Story

TLC RANCH, LAS LOMAS, CALIFORNIA

We have certainly met plenty of farmers and ranchers who have used both traditional and unconventional methods to raise capital. But one thing that unites most farmers is that they tend to be a bit private about financial matters. I completely understand the desire to keep those details concealed, but at the same time I hope we can all share more of our successes and failures obtaining financing so perhaps other farmers will not suffer as much for lack of funding. Because I know our personal farming story so well, I will share a timeline of our efforts to raise both operating and asset capital, as well as some of our cautionary tales.

Our first strategy when getting our farm (TLC Ranch) launched was what around 70% of American farmers do as well—maintain a source of off-farm income. In our case I was the one who worked full time off the farm while my husband dedicated himself to getting the farm started. We did not do this because of any socially devised gender roles, but for the simple reason that I had a stable, well-paying job with full benefits for the family, doing something I was passionate about (training beginning farmers), whereas my husband has always been the independent, self-employed type. We wanted to start a farm together, but he also wanted to develop a means of providing full-time employment for himself. With my steady paycheck we covered all of our living expenses, and we also grew much of our own food and tried to live a frugal life so we had extra to devote to startup expenses for our farm. Over the course of a few years, the farm business would end up contributing to half of those living expenses on a regular basis.

Even with living expenses covered, we still did not have any savings to speak of to fund startup costs for our livestock and poultry operation: things like buying chicks; buying pallets of feed, lumber, and other materials to build our mobile shelters; or even buying a farm truck. So we turned to 0% and low-interest credit cards. Although this is less of an option with today's credit crunch, there are still cards available with introductory rates under 10%. Once the special rates expired and shifted to a higher rate, we would transfer the remaining balance to a new card with a zero percent introductory offer. We did this for about three years until we no longer needed credit cards and didn't want to open any more accounts, which could negatively impact our credit score. Even though our credit limits were high, we never put more than $5,000 a year on the credit cards, an amount that we felt we could cover with our business sales. There is undoubtedly the temptation and the ability to go hog wild with credit cards, but we always used restraint.

Also, about midway through our very first year, a seasoned organic farmer in the county found out about what we were doing and enthusiastically latched onto us as a sort of self-appointed mentor. This gentleman gave us all sorts of ideas, stacks of poultry articles, and a $10,000 private loan, payable "whenever," he said, with no interest. We paid him back in about 18 months and were enormously grateful for the confidence he put in us. We were a regular feature at his holiday potlucks and parties after that.

Our next strategy was to begin collaborating with a well-established, large CSA operation. They needed an egg producer, and we wanted a dependable outlet for a certain volume of eggs and an up-front operating loan. Because the nature of CSA is prepayment for a share in the farm's harvest, this particular farm was collecting tens of thousands of dollars in the winter from customers. Thus they were willing to give us an operating loan in the winter when our cash flow was tight in exchange for a production contract for our eggs at a set price. We paid back the loan once our egg season began, accepting half price for our eggs until the loan was paid off, after which we would receive full price. For three seasons we accepted between $10,000 and $15,000 in advance payments from this CSA, money we employed almost exclusively to buy and raise the replacement pullets that we used to supply eggs for the CSA.

The advantages of this arrangement were an interest-free loan, minimal paperwork, and a predictable, guaranteed sales venue for our eggs. The disadvantages (these are important, so consider them seriously) were being locked into a price over the term of the contract, regardless of how our input costs fluctuated over that time; having to provide a certain volume regardless of the vagaries of weather, flock health, or theft; and getting paid half price while we were paying off the yearly loan, which was hard on our cash flow (the animals don't eat half as much even when we were paid half!). Also, even if we had more lucrative direct markets, the contract holder's guaranteed volume had to come first. We often had to short our farmers' markets of eggs to provide our weekly allotment of eggs to the CSA. That made for many disappointed farmers' market customers (people would get downright angry) and thousands of

TLC Ranch's armada of low-cost, portable hen houses

dollars in lost revenues (because our retail farmers' market price was more than our wholesale CSA price).

It is important to note that with egg production you can't ever really predict what your exact volume will be, nor when your hens will go into their molting phase and stop laying. You can estimate, of course, but it can vary as much as 20%, plus or minus, from that estimate. Also, unlike vegetables, of which you can plant an extra row or two and till them under if you have to, you can't really afford to overproduce eggs or raise a lot more hens than you need to.

Four years into our business, we decided to offer what we coined "Egg Shares," which were essentially customer prepayments for a set quantity of eggs. We started this program when we needed a large and somewhat quick infusion of capital to buy a refrigerated delivery vehicle. We had already taken out auto loans for personal vehicles in the past and didn't want to do that again, to protect our credit record; also, we wanted to avoid a little interest. Customers pre-paid for 100 dozen eggs that they could pick up either at our farmers' markets or off the farm for a price that was 15% off the retail price. Likewise, those customers always came first when we had a limited number of eggs to bring to market. We would actually hide a cooler of eggs under the table just for Egg Share customers. Customers could pick up as often as they needed to, but we capped each pickup at five dozen at a time so we had enough for other shareholders. We sold 50 of these shares and raised $25,000 in less than three weeks.

What we learned from this experience was that you need to put a limit on how many prepurchases you sell because it can squeeze cash flow later on and lock up a good amount of your production volume. Also, make certain the timeline is not too long because you don't want to be locked in at a certain price for your products when your costs of production are significantly increasing. This happened when our egg shareholders had contracted at a set price, even though we needed to raise the retail price by more than $2 over the two years of their contracts. What started as a 15% discount ended as a 37.5% discount—good for them, but not so good for us. Despite that deep discount, another important benefit of having shareholders was that they inevitably bought other products we had at our farmers' market stand and became quite committed partners in our business.

In our fifth year we decided to try to buy a home base for family living, egg washing and storage, and meat storage, but we needed a down payment to qualify for a mortgage. This time we e-mailed our customer list, which we had built up to around 2,000 people. We asked for three people to step forward and each make a $10,000 loan, payable over three years, with interest paid as a $500 credit in meat and eggs (equivalent to a 5% interest rate, more or less). We actually had five people volunteer but turned down two because we couldn't cash-flow $50,000 in private loans, just $30,000 (due to the short payback period, monthly payments were high). Also, for the house we were attempting to buy, $30,000 should have been sufficient. It turns out the house (which we had already been renting for nearly two years) was not fundable with a mortgage because of its lack of a legally approved water source.

To make a long story short, we decided to invest that $30,000 into more livestock, breeding stock, a flatbed truck with a liftgate for food-waste hauling for our pigs, and a new (used) farm truck. We signed legally binding promissory notes with each private lender, negotiated the payback length, then provided the meat and egg credit to them as they requested it over time. All three private lenders were paid off when we sold off the business in 2010. These three individuals believed in us, enjoyed the food we produced, and were looking to invest some of their money into beneficial businesses within the community—real "Slow Money" lenders.

People like this are everywhere, especially as more and more folks find out about the destructive businesses that their investments fund and also just how volatile those industries can be. People lost billions, maybe even trillions when the stock market crashed. Maybe they would rather shift their money somewhere closer, somewhere they can see, feel, and taste, and earn interest in the form of food or other essentials.

Finally, it's important to share how we, in a way, paid our gifts forward to other food and farming entrepreneurs. For two of the businesses we sold pigs to, we gave them 6 to 12 months to pay us back, interest free. We sold our laying hen flock and egg infrastructure to another beginning farming couple and gave them 12 months to pay it back, interest free. We also provided the young farmers ongoing technical assistance over the course of the year. For yet another established farmer, we sold her our farm truck and put off her repayments until the height of her production season, again interest free. Just recently, we also made two small operating loans to beginning farmers with good work ethics and business savvy.

We hope to be able to continue lending to other beginning farmers over time.

People believed in us; now it is time for us to believe in others. The banks aren't lending much, but private citizens are. It is my hope eventually to create some sort of localized, farmer-led revolving loan fund that takes on a couple of new farmers each year, with the interest payments growing the fund over time.

SELF-FINANCING?

What about self-financing? Is it possible for a business to survive and thrive just on its own profits? Well, many farmers we've talked to do just this and seem to be doing well. Unless you have aspirations of fast growth or vast scale, self-funding your growth will teach you patience and help avoid risk. It will also probably engender a more human-scale business model that will be more sustainable over time. The linchpin to make this concept work is that you must price your products right from the beginning and build a profit margin into your sales. If not, you will only be covering your *current* costs of production and won't be able to save or carry profits over into the next year to fund new productive assets. If you want to be in business for the long term, start saving some of your profits now.

If you are too tempted to spend your profits and are generally bad at saving money, consider locking them into term CDs, socially responsible mutual funds (make sure you can get your money out within one to five years), or perhaps investments in your own farm that have a strong return on investment (ROI). Some farmers, to avoid income taxes on any net profits at the end of the calendar year, will instead use those profits to secure inputs or equipment for the following year. That sounds like a wise decision to me, and it is a way to invest in your future growth.

TAKE-HOME MESSAGES

- Try not to pay too high an interest rate on a mortgage, loan, or line of credit. This will eat into what little profit margins you make as a farmer.
- When it comes to repaying debt, you need to understand your cash flows and ability to pay it over the course of the year. For

example, if you structure all your debt as short-term debt but use it for more expensive, longer-term assets, you will be paying much larger debt payments. This may eat too heavily into your monthly operating income and severely limit your operations. In addition, debt repayment in the months when you have little to no production may be difficult, if not impossible, to pay.

- Don't use up all your savings, equity, retirement funds, and leave yourself with nothing for a rainy day or an emergency.
- Don't pay more for your farmland than the production off that land will ever be able to pay for, essentially forcing you into a life of debt and probably off-farm employment.
- Don't give up too much of your meager profits to your investors, promising away the business that you are building or forcing you to sell off before you want to.
- Don't accept too much prepayment and hurt your cash flow later. I would say a good rule of thumb is not to sell more than 50% of your total production ahead of time, even if you are a CSA farm. Or have people pay you in quarterly installments so you have income throughout the year.
- Don't accept prepayment or contracting without reevaluating your costs of production each year. You might need to factor in higher input costs, especially if you plan to do any multiyear contracting. Before you sign a contract with a store, for example, you may want to build in a higher price for out-of-season product (for example, eggs in wintertime) or a percentage for input inflation. Likewise, make sure you factor in seasonality when it comes to the contracted volume (egg production will drop to 40 to 50% in the winter, typically, without all-day lighting; animals and vegetables grow much slower in the winter, etc.).
- Make sure you deliver on a contract, prepayment, and so on. Also, pay back your private investors, or you could end up in some legal trouble (also, you are creating bad karma).
- Other creative funding models typically require social media savvy, and considerable time and attention devoted to the campaign. Also, you have to deliver on the promises you make (such as with Kickstarter, where you have to promise certain "gifts" for your donors). Make sure you deliver!

CHAPTER 5

FARM PLANNING FOR SUCCESS

So NOW THAT YOU'VE GOT your training (chapter 1), identified your market (chapter 2), secured a piece of land (chapter 3), and raised the cash you need to start or expand (chapter 4), just *how* exactly will you run your business? Ideally, a thorough **planning process** will be completed before you undertake any of those actions above, but often entrepreneurs need to get started with *something* before they can realistically map out the course of their business for the future. Planning questions to consider include:

- What will your business look like, and what mix of enterprises will you choose?
- Will you raise row crops, field crops, perennials, animals, or a mix?
- Will you be expanding into value-added processing—drying, milling, grinding, smoking, cooking, and packaging—with some of what you grow?
- Will you be vertically integrating or partnering with other producers to expand your markets?

Learning how to plan for the future of your business and how to make good decisions are keys to creating a successful, sustainable farm.

STRATEGIC PLANNING

Strategic planning is frequently thought of as a process for nonprofit organizations or big business, but all businesses regardless of size should partake in some long-range planning. Strategic planning is used to identify and assess alternative business strategies and is done *before* the business planning. Business planning is used to implement a business strategy or series of strategies that are identified *through* the strategic planning process. A strategic plan is a "living" document that changes as your goals and resources evolve over time. You should aim to update this plan every three to five years or so and update the business plan portion of it annually.

The first phase of strategic planning involves creating your vision and goals, then scanning the internal and external environment. A goal-setting process that I believe fits well with creating a sustainable farming business is that of holistic management, sometimes referred to as whole-farm planning.

WHOLE-FARM PLANNING

Instead of teaching you how to write a standard, and somewhat dry, business plan, let's focus instead on helping you determine a whole-farm plan. There are plenty of business planning templates available on the Internet, but whole-farm planning is harder to find good information about. Whole-farm planning should inform your strategic plan and should be part of the process, but they are not one and the same. A handy little guide that I gleaned much of this information from is titled *Whole-Farm Planning: Ecological Imperatives, Personal Values, and Economies*, written by Elizabeth Henderson and Karl North as part of NOFA's Organic Principles and Practices Handbook Series published by Chelsea Green Publishing. Many of these concepts also derive from the Allan Savory bible, *Holistic Management: A New Framework for Decision Making.*

A whole-farm plan is a living document that includes your holistic goal, forms of production, and understanding of the ecological, social, and economic processes that impact you and your farm (situation analysis), plus a testing framework to evaluate your actions and decisions, track the results, and update your goal regularly. You will notice that whole-farm planning circles back to goal setting, demonstrating that this process is constantly evolving and adaptive to change.

First off, what is a holistic goal, and how do you form one? A holistic goal is a long-term, overall blueprint for what you want to create in your life, your business, your land, and your community. You define the parameters—what is known as the "whole under management"; essentially, what you want to manage. One strategy to form a holistic goal is to break it into three parts: **quality of life**, **forms of production**, and **future resource base**. All members of your farm's decision-making team should participate in this process. However, it may be helpful to each write out your goals individually, then come together to discuss them and identify commonalities that you can then fashion into your group holistic goal.

Part 1 of Holistic Goal: Quality of Life

Come up with 10 to 20 short phrases that describe the quality of life you want for yourself and your family. I wrote down a few examples in the table below from some recent goal setting that I participated in. Draw a table for yourself and fill it in. There is no right or wrong answer—pick the quality-of-life aspects that resonate the most with you. Then, once you have written down the phrases, describe what that phrase will provide for your life in the adjacent column, under "What Will That Give Me?"

QUALITY-OF-LIFE PHRASE	WHAT WILL THAT GIVE ME?
1. Health and wellness (for me and my family)	Good health, energy, self-confidence, security
2. Meaningful and impactful work	Feelings of effectiveness, daily joy, appreciation, contributing to a better world
3. Small environmental footprint, living ecologically	Harmony with natural world, contentment, contributing to a better world
4. Positive, close, and reliable social networks	Fun, connection, companionship, joy
5. Empathy and compassion for others	Cooperation, building culture of respect, connection, friendship

Part 2 of Holistic Goal: Forms of Production

To manifest the quality of life that you described, you probably need to do certain things in your life to get there. These are called forms of production, although they don't have to be tangible products per se. They can be personal changes, social and cultural things, or economic forms of production. Again, using the examples I provided above, here are some forms of production for the quality-of-life characteristics I am looking for:

QUALITY-OF-LIFE PHRASE	WHAT DO I HAVE TO PRODUCE OR CREATE TO ACHIEVE THIS?
1. Health and wellness (for me and my family)	Time for self, healthy cooking, good ingredients, dedication to exercise
2. Meaningful and impactful work	Write successful grants, design and implement good programs, build lasting partnerships
3. Small environmental footprint, living ecologically	Conserve energy and water, use renewables, avoid driving, restore habitats
4. Positive, close, and reliable social networks	Connect with like-minded people, find and make friends, communicate often
5. Empathy and compassion for others	Commit to service, create more love in my heart, practice nonviolent communication

Part 3 of Holistic Goal: Future Resource Base

What do you, your landscape, and your community need to look like to support the quality of life and forms of production you identified above? Answer the following questions to help paint that picture of the future.

- *What must you be like in the future?* For me I wrote "Patient, caring, empathetic, hard-working, dedicated, effective, fun."
- *What must the land be like in the future to support the quality of life you desire?* For me I wrote: "Fertile soil, stable climate, clean air, good water, open spaces, affordable farmland, and limited urban sprawl."
- *What must your community be like in the future to support the quality of life you desire?* I wrote: "Oriented toward sustainability, health and wellness, caring, and creating dignity and compassion for others; loving kids; being involved and proactive; supporting local businesses and local solutions; skilled; economically diverse; and land based.

Once you have developed each part of your holistic goal, see if you can craft it into a single paragraph or list of bullet points that you then utilize to make decisions. I should point out that since I am not running my own farm at this point in my career, my holistic goal does not include specific business goals or forms of production. Instead, it is more generally about the kind of life I want to live. A draft version of my current holistic goal looks like this:

> I want personal health and wellness as well as a healthy family. I want to be involved in meaningful and impactful work both at home and in my career, and to be appreciated for my efforts. I want to live a life with a small environmental footprint and actively restore damaged ecosystems. I want to have positive social networks with close and reliable friends for my whole family and me. I want to increase my empathy and compassion for others and be part of a family that excels at this. To achieve these things, I will make time for myself and my family, commit to healthy cooking and exercise, design and implement good work, actively design an environmentally restorative homestead to live and work within, connect with like-minded communities, and commit to service and compassion. I will remain patient, empathetic, hardworking, and fun. Our land and environment will be fertile, increasing in organic matter, with clean water, air, intact habitats, and biodiversity while still supporting working landscapes and land-based livelihoods. Our community will be oriented toward sustainability, creating dignity and compassion, health and wellness, nurturing children and the elderly, and supporting local business and local solutions.

Now see if you can shorten this paragraph into one sentence that can serve as your business's **vision statement**, which forms the base of your strategic plan. This is often one of the hardest parts of this process, but you need a vision statement that you can say to anybody in less than 30 seconds. Shortening my holistic goal above into a vision statement might read: *My family and I will create and support health and wellness while pursuing our passions, caring for our community, and improving our environment.*

Now that you have your overall holistic goal and vision statement, you can move on to analyzing the internal and external economic, social, and environmental factors that will influence your business. Once we have scanned all of those factors, we will define those that have the most influence through a process known as **Strengths, Weaknesses, Opportunities, and Challenges (SWOC) analysis**.

SITUATION ANALYSIS

What is SWOC? It is mapping out your *strengths, weaknesses, opportunities,* and *challenges*. In strategic planning this process should be done for the business *as a whole,* and for each individual enterprise you are already doing or considering. Strengths and weaknesses are usually thought of as *internal*, meaning you and your management team. Opportunities and challenges are usually thought of as *external,* meaning things happening in your environment, community, political system, and so on. Using the following grid format, try doing this exercise below. Use your current state of affairs, whether you are fully in business or just considering a farming business at this point. Include all the owners and partners in this process, because you want to write down everything that applies to your business team. Approach this exercise as you would a brainstorming session. Write down all the ideas your group can think of in 10 or 20 minutes, making sure that no idea is shunned or turned down. This will help you do a more thorough job of finding all potential factors influencing your business viability. Here is a short SWOC example:

STRENGTHS	**WEAKNESSES/LIMITATIONS**
S1. 10 years' experience as graphic designer	W1. Bad back
S2. Highly social, lots of friends	W2. High-altitude growing conditions
OPPORTUNITIES	**CHALLENGES**
O1. Well-traveled road frontage that might work for farmstand	C1. Proposed Whole Foods in town might draw away business
O2. Lots of tourists and second-home owners visit during summer growing season	C2. Summer drought getting worse

If you were going to interpret this sample SWOC analysis as your own, you would see that your graphic design background and social nature are helpful components to your business. You will be able to develop good-looking marketing materials and attract customers with your friendly demeanor. **You can build a successful business on your strengths.** If you are a social person, but your back is not good, for example, perhaps you should concentrate more on the marketing end of the business and find others that can do more of the physical production work; you don't want to make your back worse by overworking it. If you are growing in a high-altitude location, the season may be short but the solar power is intense. Maximize that short season through greenhouses or hoophouses that can also withstand the wind and snow that likely accompanies your high altitude; maybe raise some animals that thrive in those conditions or that can be grown out in a relatively short four- to five-month period. Goats and sheep come to mind, along with a summer crop of poultry or rabbits. **Look for ways to minimize the impact of weaknesses on your farm business.**

If a drought is getting worse, look into developing alternative sources of irrigation. Perhaps that could include roof-water catchment that captures the water sheeting off your buildings in the winter that can be stored for summer irrigation. Maybe you can capture snow in swales that are laid out on the contour of your land, then plant fruit trees along those swales that will utilize the melted snow for irrigation (of course, pick fruit trees that like the high-altitude conditions; peaches and apples would be a good choice). You could also make the switch to drip irrigation and enjoy the benefit of using much less water in your irrigation. Find and plant more drought-resistant varieties, too. If you have opportunities to build a farmstand, consider constructing something you could erect for the summer season when you have crops and when there are a lot of vacationers in the area looking to buy food.

If a new Whole Foods is rumored to be opening in town, maybe you can build loyalty for your farmstand now so people don't switch over to Whole Foods for produce when they do open. You could also position yourself to sell wholesale to Whole Foods in addition to at your farmstand, so you can have your foot in both markets. Having more than one market improves your resilience to changes in the marketplace. **Think about ways to turn challenges into opportunities.** Making plans and implementing actions to maximize your advantages and respond proactively to your disadvantages will prepare you better for change and competition.

Do this SWOC process annually, as situations can change dramatically over the course of just one year.

Now that you have created your vision and goals, and have taken stock of the internal and external environment, you need to determine your business strategies.

DEVELOPING STRATEGIES

A strategy is the means by which the business uses its strengths to take advantage of opportunities so the goals identified for the business can be realized. If you have already identified your strengths and opportunities doing the SWOC analysis above, you are halfway there. However, this is often the most difficult part of strategic planning because you have to create and evaluate your different options. One way to do this is first to identify two or more future scenarios for your business. This could involve different enterprises or various ways of using resources. From these strategic alternatives, the alternative that best achieves your holistic goal is chosen. Some common farm business strategies with examples include:

- **Growth**—Expanding the size of the business in some way. For example, you will contract with nearby chicken farmers to grow pastured broilers using a defined management protocol to increase your production without stressing your resource base.
- **Stability**—Maintaining the size of the business. For example, you will continue to refine your production practices so you cut costs and improve quality to maintain your existing markets.
- **Retrenchment/downsizing**—Refocusing the business for improved performance. For example, you will be dropping your unprofitable hay enterprise and replacing it with dry bean production.
- **Succession**—Transferring the business to a younger generation. For example, you will be separating the business from the farmland assets and creating a new family-owned LLC for the business to facilitate transfer.
- **Exit**—Ending and leaving the business. For example, you will be organizing your last five years of financials so you can attract a buyer for your business.

Once you have identified a range of scenarios, you have to devise a method to pick which strategies to pursue. To test the different scenarios against your holistic goal, start by asking yourself the following questions:

- Does this decision move the business toward or away from my three-part holistic goal?
- Does this action maximize my strengths?
- Does this action minimize a weakness and get to the root cause or the weakest link?
- Does this action turn a challenge into an opportunity?

You can also use these four questions above to test your day-to-day decisions to make sure they are in line with your strategic plan and holistic goal. If this seems like a lot of work to make day-to-day decisions, it might be at first, but you will soon get used to running through these criteria in your head instinctively.

How do you find the weakest links in your farming model? Essentially, you have three relationship chains that you are managing as a farmer: **the social chain, the biological chain,** and **the financial chain**. Go through each one on paper by drawing them out, showing directional arrows, and writing down any challenges or problems you may be having at each step of the chain. Here is an example:

EXAMPLE OF SOCIAL CHAIN WITH CURRENT WEAK LINKS

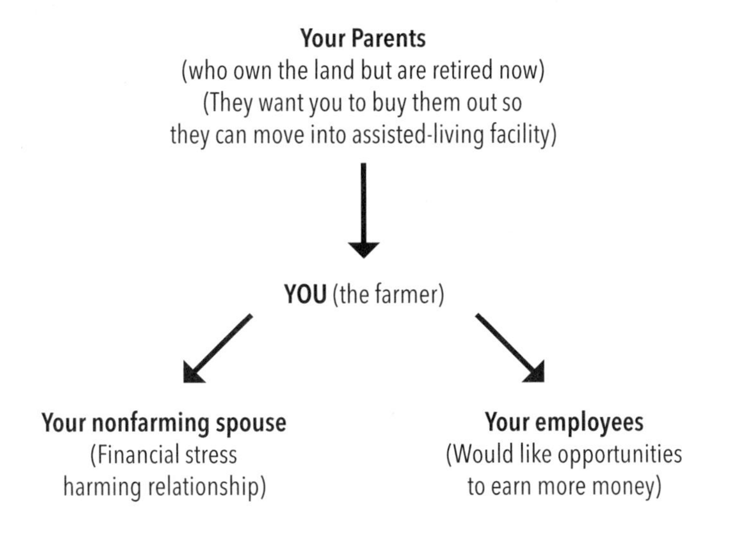

In this example your current social chain weak links are your parents wanting to get cashed out, your relationship with your spouse weakened because of financial stress, and your employees, who would like to earn more remuneration. As you devise potential solutions to these weak links, ask yourself: Does this action deal with any confusion, anger, or opposition it could create in people whose support is needed in the near or distant future? How could I enhance this action to further strengthen relationships, collaboration, enjoyment, and so on? Can I take small steps to address all three weak links instead of prioritizing one over the others? For example, you could look into selling the development rights to your farm so you could pay off your parents while retaining land ownership; take a holistic financial planning course with your spouse so you can begin to form financial goals together and speak the same language; and implement a profit-sharing program with your full-time employees so that they will do better as the business does better.

Next, will your business strategies and forms of production focus on producing high volumes of products (quantity) or on producing high-quality foods? This is an essential question to ask yourself when developing your strategic plan.

FOCUS ON QUANTITY OR QUALITY?

Our food system has spent a great deal of time and resources in the past 70 years focused on volume and cost cutting. As a farmer, of course you should be attempting to maximize volume, *given* your resources (social, environmental, financial). Being efficient with what you've got makes good sense all around. But do so with an observant eye toward the carrying capacity of your soils, water, native vegetation, wildlife, labor, transportation system, energy consumption, and more. You should also constantly be paying attention to your costs of production and your overhead, devising ways to reduce those costs while staying in line with your values and holistic goal. But as the small or midscale farmer that you probably are or strive to be, can you be the least-cost, high-volume producer?

The answer is probably not. When I see farmers attempting this, I often see failure or bankruptcy result. For example, there is a tiny 1.5-acre urban chicken farm I know of in a progressive West Coast city. They have scaled up their laying hen numbers to 3,000 birds, and they wholesale their eggs to retailers all over town. Not only has the natural resource base been thoroughly degraded (no more pasture for these supposedly pastured hens), the hens suffer from more disease pressure, and the farmer makes so little money wholesaling that he has to work full time off the farm. How might his resource base, flock health, and profitability change if he changed his business strategy by downsizing to 800 birds and selling his eggs for twice as much at farmers' markets?

On the flip side there might be enterprises that you could scale up a bit to realize some economies of scale, and maybe you'd be able to pass *some* of those savings on to your customers. For example, if your state requires you to build a licensed dairy facility whether or not you are milking 2 cows or 20, maybe you should milk 20. Likewise, so much of the dairy equipment available is for larger-scale facilities, so maybe you need a few more animals in production to fully utilize the equipment you purchased and spread those fixed costs over more animal "units." My long-winded point is that your niche as a farmer will probably not be that of the cheap, go-to supplier of a commodity. So what, then, will be your defining role as a food producer? Might I suggest **quality**?

When people think of quality foods (or fibers, herbs, whatever you grow), they think of freshness; location; products grown in clean soil, in clean air, and with clean water that are minimally processed or adulterated, flavorful, tasty, juicy, aesthetically pleasing, nutrient dense, well made, last longer, packaged well, from a respectable producer, and priced accordingly. How will you compete on quality? We will discuss this at more length in chapter 9, marketing. However, the key is that you value quality yourself and plan for quality in your strategic plan and holistic goal. Then work to maximize those qualities of your food products as you gain experience over time, and make sure you convey that commitment to your audience (customers), whomever they may be.

Once you determine your business strategies, you should then use them to create a business plan, a human resources plan (discussed in chapter 12), and a marketing plan (chapter 9). But remember, plans are only as good as their implementation and control.

MANAGING CHANGE

As your farming business matures, you will likely add new enterprises, drop others, explore new markets, acquire new productive assets, and, I hope, learn from your mistakes. All of this means **change** is on the horizon. How do you plan for that change and create a flexible business model? Here are some potential change management steps.

- First, plan to revisit your holistic goal and strategic plan probably once a year. How about taking one or two days in early January to sit down with your team to do this?
- If you have a business plan, too, try to revise this once a year as well, right after you update your holistic goal and strategic plan. They all have to be in line with each other.
- Review your financial statements to identify weak links and potential areas for "cost containment," and run a gross-profit analysis of all your enterprises.
- Plan for new enterprises through budgeting, cash-flow analysis, and compatibility with other enterprises, and examine how you can improve your resource base (natural, social, and financial).
- Discuss areas that are causing dissatisfaction, stress, pain, safety hazards, or negative impacts to your resource base. Talk about why these problems are happening (go as far back as possible, to

the *source* of the problem if you can), and discuss plans for solving those problems.
- Test all your decision making against your holistic goal.
- Plan for some redundancy of labor, in case people get sick, are injured, or quit.
- Get further training in areas that provide more flexibility and enhance existing skill sets (your own and those of others).
- Borrow or rent equipment until you absolutely need to buy it.
- Buy equipment for the scale you plan to be in 5 to 10 years.
- Don't overbuild or overbuy for new enterprises you are just testing out or ones that are temporary or short term.
- Buy low, sell high. I will expand upon this concept in chapter 6, the equipment chapter, but I mean that you need to make affordable, strategic purchases of equipment and other productive assets that you could then turn around and sell for the same price, or perhaps more, if you decide to change enterprises, scale up, and so forth.
- Create exit plans for dropping enterprises, ending the business, going into retirement, or passing on the farm to the next generation.

Although most of the farm stories I have included in this book contain elements of good planning, the following story of one Massachusetts farming couple displays how planning, implementation, and adaptation all must work in tandem to create a lasting farm business. Misty Brook still has its struggles, particularly around land tenure, but they are building a resilient, sustainable business in spite of that challenge.

Goodness from the Ground Up

MISTY BROOK FARM, HARDWICK, MASSACHUSETTS

• Katia and Brendan Holmes •

As first-generation farmers in an expensive state with dwindling affordable farmland, Katia and Brendan Holmes seem like they have a steep hill to climb with no clear view of the top. But their Massachusetts-based farming business, Misty Brook Farm, is a lesson in tenacity, strategy, and utter devotion to values. Despite their struggles they are making it work for them and would be considered successful if one only looked at their income statement. Their success, however, is much broader than that.

Brendan knew as a child that he wanted to be a farmer and thus spent his summers and afternoons working on nearby farms in North Yorkshire, England, then later in Massachusetts when his family moved there. Katia, another suburban Massachusetts kid, thought farming looked like more of a hobby than a bonafide way to earn a living, so she decided to pursue a career as a large-animal veterinarian. The couple met while employed at Seven Stars Farm in Pennsylvania, both trying to hone their skills with livestock. After a few other farming jobs, some in which their ideas were not always heard, they left for an intensive three-year education in biodynamic agriculture at Emerson College in England. There they learned not only the principles of biodynamic and organic production but also valuable business management skills that would serve them well once they started their own farm.

Even while working for other farmers (including organic dairy pioneers Jack and Anne Lazor of Butterworks Farm in Vermont; see their narrative profile in chapter 12), Katia and Brendan dabbled in producing their own food and selling the extras to friends and neighbors. They always had some poultry, a vegetable garden, and sometimes a few pigs they would sell a side of. This is how they first got exposed to direct marketing, and from there on out they decided that selling straight to the consumer was the way to market the fruits of their labor. The couple liked the direct consumer connection in terms of building community, shortening the food chain, and earning a better living. Now, six years into farming on their own, they are nourishing hundreds of families with nearly everything a person needs for a healthy diet.

The Misty Brook farming venture began with little money; both Katia and Brendan had student loans to pay off and no real savings. They worked odd jobs while getting started, sold firewood, and used no-interest credit cards to finance their initial startup. Although it is no longer the case these days, credit cards used to offer promotions of 0% interest for 12 to 18 months at a time. Before the card would switch to the double-digit interest rate, the Holmeses would move the balance to another 0% card. They eventually paid off that debt with the farm sales, all the while paying very little interest. For young farmers with little credit history and no real opportunities for traditional bank financing, this is a good way to go as long as you show restraint and stay on top of the interest rate changes, which the Holmeses did.

At first the couple started out by selling sides of beef, buying either yearling stockers or weanlings and finishing them for a year or so on grass. This enterprise was not very capital intensive and had quicker payback than raising beef cattle for the full two years of their lives. It also helped them develop a market and improve their pasture-management skills before committing to the full calving-to-finishing operation. After this, the Holmeses started purchasing a few dairy cows and heifer calves to expand their dairy operation and would produce milk-fed veal with any bull calves born. However, they didn't receive their raw-milk license from the state after a somewhat lengthy process, so they began raising pigs on the milk they could not yet sell. Pigs are a great complementary enterprise with any dairy because they can help balance out any excess milk supply or drink the milk that does not meet a farmer's quality standards for whatever reason.

Katia and Brendan also started raising broiler chickens, ducks, and laying hens, then added mixed vegetables and grains to their diversified farm. They even dabbled in sheep but realized sheep were not complementary to their other enterprises, eating too much of the grass needed for the high forage requirements of their dairy cows, and also too labor intensive, mainly because of the increased fencing they require. To accommodate all of their enterprises, plus the forage and feed they must grow for their livestock, the Holmeses must lease 13 or 14 different parcels of land within a 20-mile radius of their home in Hardwick. By growing their own feed, they can reduce their risks of a widely fluctuating global grain market as well as control for quality, making sure they get the freshest, most nutrient-dense grains and forages for their animals.

Their land tenure situation is the biggest challenge for them and one that has nearly put them out of business before. Katia and Brendan are still recovering financially from a quick move they had to make a few years back when their main farm lease was terminated for no legitimate reason. Their current lease agreement on the dairy farm is only month to month. Despite investing tens of thousands of dollars into the dairy infrastructure and building a farm store, getting the land certified organic, and revitalizing all of the pastures, they foresee it will be only a couple of years more until their landlord decides to move his own cows back into the dairy barn, and Misty Brook will have to move once again. They are hoping their next move will be their final one.

Likewise, the cost in time and fuel to cultivate forage and grains and move animals between different fields they rent is probably equivalent to a full-time employee. It is not only hard on their economic bottom line but also places some burden on their family life. They dream of the day when they can walk out their front door to milk the cows while their sons can keep sleeping in their warm beds.

Trying to manage all these enterprises could easily mean 20-hour workdays for both Katia and Brendan. They employ various strategies to manage their workload and keep their labor costs down. In the summer, when milk consumption goes down because schoolkids are on summer break and families often on vacation, the Holmeses convert their dairy herd to once-a-day milking. They breed their cows to calve in the fall so milk production spikes when consumer demand also goes up. Once-a-day milking allows them to focus more of their labor on the demands of growing and putting up forage for the winter, growing vegetables for their customers, and getting home a little earlier at night to spend more quality time with their family. The couple also rotate the daily milking chores, which allows one or the other to sleep in with their

KATIA HOLMES

The symbiotic relationship between a farm family and their cows

young sons and not burn out from the seven-day-a-week milking commitment.

They have a few employees: One helps with all the tractor work and maintenance, one works part time bottling the milk and sterilizing the returnable bottles, and two work part time growing and harvesting the vegetables and organizing CSA pickup day. They have offered seasonal internships for the last few years, providing housing and a modest stipend in exchange for full-time work. They strive to be a good employer, but the high cost of living in the region makes it hard to provide living wage salaries. To augment the modest wages, they provide food and flexibility of hours. Several of their employees have been with them for years and are treated like part of an extended family. Katia and Brendan want their farm to be an economic driver and job creator. They envision a time in which a dozen or so people are either employed directly or creating their own small businesses as part of Misty Brook Farm. They would love to see a couple take over the vegetable production, for example, or bake bread for the farm's CSA and farm shop.

Katia and Brendan market virtually all of their products through their self-serve farm shop or through their various CSA subscriptions, a business strategy that has worked quite well for them. Only on rare occasions do they sell to local restaurants or caterers, usually a whole or half butchered animal. The farm shop is completely on the honor system, accepting cash or check in a locked money box (exact change required). CSA customers are encouraged to pick up on

Fridays when a large assortment of fresh vegetables is harvested and put out for choosing, but they can also elect to pick up any day of the week directly from what is in the farm shop. Misty Brook offers three unique CSA options:

Whole Diet CSA: Year-round; members pay a set price per person, which allows them to take whatever food they want, as much as they need, either from the farm shop or on Friday pickup days; based on the value of $10 per person per day. They have about 15 families in this program.

Meat CSA: In 2011 the price was $550 for four months and $800 for six months. Members get to take a $160 value from the freezers each month, so they get a discount on the meat through the share. Meat options include beef, veal, pork, and chicken. Customers get to choose which cuts they want and how much they take at one time.

Vegetable CSA: Seasonal; members pay $650 for 22 weeks, and they get an organic cotton bag, which they fill up each week with any vegetables they choose, plus other produce they want to harvest directly from the fields, such as strawberries and peas.

Marketing of the Misty Brook farm shop and CSA shares are mostly via word of mouth. Their customers come from the local community but also from up to about two hours away—folks who are sincerely appreciative of the high-quality organic foods that Misty Brook provides. They include working-class families, wealthier childless couples, retired people, and others who are concerned about their health, want to support local agriculture, and appreciate the quality, taste, and freshness of Misty Brook's foods. Katia and Brendan don't use social media to advertise, except for a simple website they created a little over a year ago. Free web listings with Localharvest.org and EatWild.org provided much of their initial clientele.

By focusing on a CSA, Misty Brook does not feel compelled to continue growing in scale but rather can focus more on improving quality and keeping its current customer base satisfied and loyal. CSA customers always come first—whatever is left is sold on a first-come, first-served basis at the farm shop. To keep labor costs down, produce is harvested into totes and left unwashed and unbagged. Some labor-intensive vegetables, such as peas and strawberries, are left as pick-your-own crops and not harvested for the customers. Likewise, if there are surpluses of certain crops, customers are encouraged to walk out into the fields to harvest as much as they want.

When Katia and Brendan first started milking their own cows, they fed a little organic grain during the milking. When customers started asking for 100% grassfed milk, they took the idea into serious consideration. Once they tried it, they were sold on the idea of a forage-fed herd. Although it requires an intense commitment to pasture management, improving herd genetics, and planting and putting up forage for the winter, it has ended up being more profitable because organic grain prices are up, and their milk quality is better than ever. I

A diversity of crops grows healthy and strong.

don't know how long a normal quart of raw-milk cream usually lasts, but the one Katia and Brendan gave us the day we left their farm lasted nearly a month in the refrigerator without going sour. The cream, quite simply, was the best I have ever had. My husband and I would fight for it in the mornings making our coffee.

Their short-duration, high-intensity grazing is part of the recipe. Letting the pasture grow to a couple of feet high and moving the cows twice a day using highly portable, fast-installing electric fencing is one of their innovations. Not only is it low capital and easy to install on leased land, but it provides for the intense forage needs of a 100% grassfed lactating cow. Katia continues to try to improve the genetics of their mostly Jersey herd, selecting the most durable cows that thrive on grazing.

Recently, the Holmeses were able to purchase a fixer-upper house in the town of Hardwick, about a 15-minute drive from the dairy. The house not only gives them a space to have an office, a phone line, and a permanent address to run their business from but also a couple of acres in which to store equipment that they are repairing, raise some chickens, have a perennial orchard, have a greenhouse for veggie starts, and provide housing for apprentices. Because they bought it cheap, they will be able to build some sweat equity in the property and perhaps use that to purchase a farm someday.

The beauty of this whole transaction was their use of unconventional financing. Banks were not lending, especially for two self-employed farmers. Instead, they asked their customers if they would be willing to make loans to buy the house. The response was so great they actually had to turn away a couple of offers. They negotiated interest rates and payment

Lightweight portable fencing makes rotational grazing a cinch.

schedules with about a dozen people, using an official promissory note for each one. In about 15 years they will own their house outright and will have paid much less interest than they would have with conventional bank financing. At the same time their lenders will be earning better interest than with most places they could have invested their money. It is truly a win-win scenario and I hope a step in the right direction to building longer-term security for their farming business.

Katia and Brendan's advice for new farmers?

- Get as much experience as you can on other farms: conventional, organic, and other variations. The wider the range of farms you work on before starting your own the better.
- Always try new things. Be creative and flexible.
- Listen to your customers, but be willing to drop unprofitable enterprises.
- Be brave enough to price your products to support yourself. Some people will try to get a bargain, but stick to your price. You need to cover all of your costs, including unexpected mishaps and future growth.
- Don't buy equipment until you really need it. Get by with what you have or borrow or lease equipment. The Holmeses didn't buy their first tractor until three years into the business. Equipment depreciates, while livestock appreciates in value.
- Get a really long-term lease on a good piece of land, at least for your core business. A good long-term lease or creative financing is your biggest asset.

TAKE-HOME MESSAGES

- Strategic planning helps you map out the long-range vision, strategies, and actions that increase the profitability, the competitiveness, and the ultimate sustainability of your business.
- Business planning is one component of a strategic plan, but they are not one and the same. Business plan templates can easily be found online; the harder process is doing the strategic plan.

- Holistic goal setting should be done with your entire ownership and management team for it to be most relevant and for there to be good buy-in.
- Instead of making decisions "by the seat of your pants," take the time to set goals and thus make decisions based on a long-term vision. If you never have time to step back from the day-to-day management to do some strategic planning, bring in some outside consultants to help give you some perspective.
- It's great to have high ideals and values in your production, but you also have to focus on efficiencies, cost controls, and running a good and efficient business.
- Analyze each enterprise individually and how they interact: Are there synergies between them, or do they compete for resources?

CHAPTER 6

EQUIPMENT AND INFRASTRUCTURE

I MIGHT BURST YOUR BUBBLE a little bit with this chapter. I know a lot of folks whose pulse accelerates every time they envision all the big, shiny green, red, and blue farm toys parked in their dream farm's sprawling equipment barn. But there's sort of a farming "law of physics" you just can't deny: The more "steel" you have, the more likely you are to use it! And the more you use it, the more fuel you will burn, the more money you'll spend on maintenance and repairs, and probably the more passes you'll make over your soil (and damage you'll do to it). On the flip side, when you don't have a lot of steel, you are forced to get creative. Case in point: Because we didn't own a tractor on our California farm, we did almost no tillage over the last six years of farming. Instead, we paid a neighbor about twice over those six years to till our fields for reseeding pasture or forage crops.

Learning to farm without a tractor, we eventually understood how to revive our pastures through animal impact and some broadcast seeding with a cheap $75 hand-cranked seeder. Our land management actually improved because of our lack of equipment, and we spent little to nothing on equipment, fuel, maintenance and repairs, or soil amendments to make up for what we would have destroyed through frequent tillage. Other than a pickup truck, the only other piece of large farm equipment we owned was a four-wheel-drive ATV for hauling feed around and collecting eggs. Likewise, we sold the ATV for the same price we paid for it after three years of steady use, so we didn't lose money to depreciation as with most farm equipment experiences.

Our "buy low, sell high" philosophy for equipment worked really well for us over the years. We did our homework and found great deals on equipment in good shape, took decent care of the equipment over the years, cleaned and fixed it up to prepare it for sale, then were often able to sell the stuff for more than we paid for it, or for the same exact price. Just as we worked to add value to the foods we sold, we did the same with our equipment, our breeding animals, and our mobile infrastructure, which I will discuss later in this chapter.

Am I suggesting that all farmers use only their backs and brawn to do all the physical labor on the farm? No! In fact, buying equipment and **not** using it is just as bad for your financial health and your farm's sustainability. I have met a lot of farmers who started out with a good deal of capital from their previous careers and proceeded to burn it up on fancy equipment that mostly sits idle. Then they complain a couple of years later that farming "just doesn't make any money." Of course it doesn't make a profit when you lock up all your available capital in superfluous equipment or infrastructure. Instead, put a lot of forethought into purchasing equipment. Here are some criteria you might use to inform your **equipment** decisions:

- Does it enhance human health and safety?
- Does it improve animal welfare?
- Does it improve any natural resource conditions?
- Does it burn fuel? If so, how much and what kind? Are there more fuel-efficient models out there that will meet your needs?
- Does it reduce labor hours for a given task, particularly repetitive tasks that are done quite frequently?
- Will it improve profitability, and over what kind of time frame?
- How much will it cost over time, including maintenance, repairs, fuels/energy, and depreciation and replacement costs?
- Will it help you increase production volume in any appreciable way?
- Will your customers view it favorably?
- In solving one problem, does it create others? Will those other problems undermine its value and utility?
- Is it built to last? Does it have a warranty? Will you be able to repair it yourself, or do you need to hire specialists for that?
- Does it help you reduce loss to pests, disease, soil erosion, drought, flooding, theft, and so forth?
- Is it in line with your holistic goal?

Likewise, before constructing anything such as irrigation systems, permanent fencing, animal shelters, corrals, barns, packing sheds, greenhouses, or walk-in freezers,consider these additional criteria for built **infrastructure**:

- What kind of energy needs does it have? Will it be able to utilize nonpolluting, homegrown, or renewable energy?
- Will it decrease climatic risks, such as the risk of drought, flooding, intense heat, prolonged cold, hurricanes, and storms?
- Will it help you improve water quality or ensure better water quantity?
- Will it appreciably reduce predation by humans, animals, or bugs?
- Will it reduce death loss (of livestock and poultry)?
- Can you build it according to the terms of your lease or the zoning of your property?
- Will it require costly permits, licenses, and engineered drawings to build?
- Will it improve relations with neighbors, your customers, your community?
- Will it enhance working conditions for your staff?
- Can you afford to build it right? If not, is there a temporary solution, or can you build it in stages as you have the money?
- Will it enhance your economic viability?
- Will it improve food safety or food quality (such as with proper cold storage)?
- Can you build it with your own skill set and labor availability, or do you have to hire specialists?

Once you have analyzed all these criteria to inform your decision, look at the pros and cons, as well as the costs to borrow, rent, or buy a piece of equipment or infrastructure; maybe set up a little decision-making table such as these below:

SAMPLE EQUIPMENT PRIORITIZATION

ITEM	BORROW/RENT/BUY?	LENGTH OF USE	PRIORITY LEVEL	YOUR BUDGET
Farm truck	Buy because of excessive wear and tear, frequent use	5 years (really beat it up)	Highest	$10,000
ATV	Buy because of excessive wear and tear, frequent use	5 years	Medium	$5,000
Egg washer	Buy because of lack of availability, frequent use	30 years	Highest	$12,000
Walk-in freezer	Rent space in nearby cold storage facility or lease freezer unit	10 years	Medium	$5,000
Seed drill	Borrow because of limited use, or rent	30 years	Low	$500

Do the same thing for prioritizing infrastructure improvements, such as permanent fencing, animal shelters, equipment sheds, housing for apprentices, or greenhouses.

SAMPLE INFRASTRUCTURE PRIORITIZATION

ITEM	DIY OR CONTRACTOR?	USEFUL LIFE	PRIORITY LEVEL	YOUR BUDGET
Permanent fencing	DIY	20 years	Highest	$500/acre
Animal shelters (10)	DIY	10 years	Medium	$500
Equipment shed	Contractor	30 years	Low	$15,000
Apprentice housing (2)	DIY if trailers; contractor if built	20 years	Low	$20,000
Greenhouses (3)	Contractor	20 years	Medium	$30,000

How might you use these tables above? You could pick off the high-priority items, compare their cost to your overall farm budget, and consider their useful life. You should also run through the criteria questions I listed above to see which projects address the most criteria: You may be surprised how your prioritization changes as a result. Also consider the financial implications, such as will you be able to finance it over the course of the useful life; say, with a bank loan or via leasing? If you can't qualify for a bank loan or find a leasing option, can you afford to pay for it all right now? Will that negatively impact cash flow and hinder operations?

Additionally, could you build some things one at a time as you have the money instead of all at once, such as with the animal shelters or greenhouses? Do you have the time to build things yourself, or do you have to bring in a contractor or tradesperson because you are too busy? (This will obviously affect the budget for the infrastructure!) You can also have contractors do the parts of the project that are too challenging for you, such as the electrical or plumbing, while you do the rest.

THE BUILDINGS AND INFRASTRUCTURE MAKE THE MANAGEMENT

The way you design the physical infrastructure of your farm or ranch will have long-term implications on your quality of life, your ability to steward your natural resources, your day-to-day operations, and your financial fitness. Living in an ideal world, I think we would all spend a year or two observing the land before we built a thing—a key concept in the world of permaculture. But we usually have to get moving on things quicker than that, such as developing water sources and building irrigation systems, animal shelters, even basic fencing to keep out predators or herbivores that want to eat all your crops. The way you build things will dictate your management for years to come; keep that in mind as you go about planning and constructing things. What may look cute and quintessential now could become a giant labor sink in the future. What may be cheap and expedient may become a safety hazard or cause natural resources degradation in the future. Some examples of how the building makes the management are illustrated below.

If you build a fancy, permanent perimeter fence, you might be tempted to do continuous grazing with your livestock, which can lead to overgrazing, parasite loading, and underutilization of some parts of your pasture. It is not certain that you will do this, but it becomes tempting and easy. If, on the other hand, you don't have a lot of money, or you don't own the land you farm, you can set up a portable polywire electric fence with a battery-powered charger. As a result, you will have a system that lends itself to rotational grazing and now—voilà!—you have improved your land management and animal husbandry possibilities. We saw that situation play out with Misty Brook Farm in Massachusetts, which was featured in the last chapter. The family doesn't own their land and thus have devised a low-cost, easy system of portable electric fencing that facilitates moving their dairy herd twice a day. It takes Brendan and Katia Holmes only 15 minutes to set up a new pasture for their cows. Their pastures, and in turn their cows, are vibrant and healthy as a result.

If capital is a limiting factor, or you are simply trying to create low-tech, low-cost solutions, you can do without many pieces of equipment by substituting human labor or harnessing animal impact. We have seen many great innovations like this in our travels that are also cost saving. One example is turning your mobile chicken tractors into winter greenhouses by taking off the tarp and putting clear plastic on top instead. I have also seen the opposite, in which a greenhouse used for summer vegetable production houses the laying hen flock over the winter. The chickens consume the leftover vegetable plants and provide fertility for the future year's crop. Another example is using pigs to rototill your weeds and fertilize your farm fields for subsequent crops; we've met several farmers doing this with success. While a rototiller costs money to fuel and maintain, a pig can be sold for a profit once it is done tilling your fields! Turn your costs into income generators whenever possible!

One young farming family that is getting their start on a limited budget is profiled below; they are using a combination of family and community resources, along with a knack for devising low-tech and low-cost solutions. Matt and Jerica Cadman are moving their nascent business along in a positive direction, adding equipment and infrastructure in a strategic way. Read on to be inspired by their vision and their actions.

It Takes a Family to Raise a Farm

SHADY GROVE RANCH, JEFFERSON, TEXAS

• Matt and Jerica Cadman •

Matt and Jerica met while studying engineering at Letourneau University in Longview, Texas, and married the summer before Jerica's junior year. This newly married couple never expected to become farmers, nor did they have any background in agriculture. But the bumpy ride of a chronic autoimmune disease forced them to take a good, hard look at their diet. Matt was diagnosed with ulcerative colitis, an "incurable" inflammatory bowel disease that seemed to be exacerbated by stress and dietary factors. At one point Matt had lost so much weight that he looked like he was fresh from a prison camp—only 120 pounds on his 6-foot 5-inch frame. They tried a couple of years of conventional medicine, hospitalizations, and drugs, to no avail.

That is when they stumbled upon the early twentieth-century research of Weston A. Price, a holistic dentist, and realized that Matt's diet might be the cause, or at least the irritant, that led to his frequent painful symptoms. Meanwhile, they were running out of options: Doctors had suggested that Matt have his colon removed. At 22 years old, he didn't think that looked like a good option for living the rest of his life to the fullest.

In their research on alternative therapies, the demise of vitamins, minerals, and healthy fats in the American diet became apparent. The Cadmans started adding grassfed beef, pastured poultry and eggs, and raw milk to their diet but had to drive all over the state to acquire these products, which were hard to find in East Texas. At the same time Matt was increasingly yearning for a home-based business so he could work hand in hand with his new wife and be around when they started having children. Starting a farm together began looking like a viable option, not only as a way to make a living together but also to provide healthy, nutrient-dense foods to their family and others who suffer from health problems.

The couple realized they needed some training in agriculture, so they applied to manage a cattle ranch, and despite their inexperience they were hired. In a year they learned quite a bit about installing fencing, managing pastures, and rotating animals, as well as a few lessons of "what not to do." Toward the end of their ranch employment, Matt's father, a retired church minister, called him up to ask how much it would cost to purchase a "starter" farm and enter a partnership with the young couple, in hopes of making their dream become a reality. The elder Cadman's stock market investments were losing money, and he thought he would rather invest his money into something real—his own son.

Matt immediately identified a half a dozen properties that had some of the characteristics they were looking for. The first one they visited was a keeper—it had a little over 100 acres, gently rolling topography, natural ponds, tree groves, a decent double-wide trailer, and a few outbuildings, and they could envision the possibilities of the new farm's taking shape.

During the tail end of Jerica's college education, once Matt and Jerica had become highly active in their pursuit of great food, they shared their desires with some fellow church members, who also caught the vision. Beginning in 2008, along with a couple of former college friends, Matt and Jerica formed a small company called TrueFields LLC. They started out with seven pigs, a small dairy herd, and a vegetable garden and kept them all under a lease arrangement with some ranchers, who would later become Matt's employers and mentors in farming. The dairy herd was used for a herd-share program, in which customers share the expenses and production of raw milk from the herd. (As a private, contractual relationship between herd owners and managers, herd share has thus far been allowed in the state of Texas, though it isn't in some other states.)

They started small, milking just two cows, but have no intentions of going beyond six to eight cows lactating at any one time. They currently have between 30 and 40 herd-share owners. If milk production ever drops, such that they can't supply all of their owners, TrueFields LLC will buy a few owners' shares back so nobody is shorted milk. This relationship with their friends as business partners has not always gone swimmingly, and a couple of years into their work, they decided to buy out the stake of one of their partners in the business. They found that as their visions developed and the business grew, their management practices, goals, and means of attainment began to diverge drastically. However, going into partnership with others at first allowed Matt and Jerica to avoid debt by not having to fully capitalize the business all by themselves.

Once Matt's father became involved with the farming dream, a couple of years after TrueFields LLC was formed, Matt and Jerica were able to relocate their current operations and establish their work on their own land. They formed another

Dairy and beef cattle enjoying their grass lunch.

company and named their land the same: Shady Grove Ranch. The farm is located on Shady Grove Road near the historic little town of Jefferson, and despite their determined efforts to find a name that might be more symbolic to them, Shady Grove Ranch was cute and catchy enough.

They continue to run TrueFields LLC as a separate business entity, keeping it separate from the farm assets to reduce their personal liability of running a raw-milk herd-share program. The Shady Grove Ranch business runs the beef herd, pastured pigs, broiler chickens, turkeys, and some laying hens. They may add meat goats and lambs in the future, as well as one day expanding their vegetable garden and maybe even planting some fruit trees to provide their customers with more of what they need. Their long-term vision is to support around 100 families with everything they need to have a healthy diet, ideally having most people pick up directly from the farm.

Marketing their meat, milk, and eggs first began with college friends and fellow churchgoers and has vastly expanded from there because of a lot of positive word of mouth. Shady Grove Ranch has a comprehensive website with regular blog updates, recipes, pictures of the farm, and even information about nutrition and health. Jerica makes it a point to attend health fairs and other community events where she can get the word out about their farm and the health benefits their food provides. Other less direct means of marketing have come from such activities as speaking at special-diet support groups and working part time at a local health food store.

Even before she started farming, she became chapter leader of her local Weston A. Price Foundation chapter, a group that emphasizes pastured and grassfed meats, healthy fats, and eating only whole, nutrient-dense foods. The Cadmans are so excited about the results they've seen in their health that they joke they can't even go on a date without meeting someone they try to "propagandize" about real food.

The Cadmans sell meat and eggs at their small, local farmers' market in Jefferson, one that Matt actually helped get off the ground, and a larger market in nearby Shreveport, Louisiana. Additionally, quite a few customers pick up in bulk at the farm, buying sides of pork or other meat to stock their freezers. Future plans include building a small farm store and maybe some sort of buying club or delivery route to Shreveport once the farmers' market closes for the season. They also hope to implement a customer-rewards program to provide savings to customers who purchase regularly and who refer other customers. The Cadmans believe word of mouth is key and to date have not hired any professional advertising.

As for the production practices of their farm, Matt and Jerica practice rotational multispecies grazing, moving their animals on a frequent basis to new pasture. They hope that someday their pastures will be able to support the cattle for at least nine months of the year without supplemented forage, and preferably more, once there is better diversity of plants, such as more legumes and cool-season grasses and increased drought resistance of the soil as more organic matter is accumulated. Running broiler chickens in movable, bottomless pens on their cattle pastures has already made an immediate improvement to the forage quality—you can see the dark green strips crossing the land where the chicken tractors were last year.

Drought, however, has not been easy on their farm, a fact driven home when a wildfire tore through their property in late summer of 2011. Fortunately, their home, animals, and infrastructure all survived, but nearly all of their pastures and trees were scorched to black carbon.

Matt and Jerica try to have True-Fields' dairy cows bred so that three cows will freshen every six months, helping to stretch out their milk supply to year-round. Hay is a huge cost for them, especially the expensive, high-protein hay they purchase when there is not enough pasture to support the high-energy needs of their dairy herd. Hay can make or break an operation, and the Cadmans are keen to figure out ways to reduce their dependency, by elongating their pasture season and plant diversity and potentially cutting and baling some of their own clover hay someday. They even aspire to grow their own pig and chicken feed in the future, realizing that their dependence on long-distance trucking and fossil fuels is not sustainable over the long term.

Not having to pay a new mortgage with the proceeds of farming has surely been a great advantage for the Cadmans. They are convinced that kind of debt load would not have been sustainable for them, nor allowed them to farm with the values they have. They have formed business relationships with several like-minded college friends who also desire to see the farm grow faster by providing some small investments to move things

JERICA CADMAN

Matt and son putting out supplemental feed after the fire.

along. Still, their primary means of capital acquisition has been marketing what they do have so that they can grow more.

They have been very slow and cautious about the investments they have made, always buying used equipment for the best price they can, then often putting a little work into it.

They were even given a free nonworking ATV and figured their engineering minds could fix whatever was wrong with it. It's low on the priority list, so it's not fixed yet, but they have made do. They purchased a used double-jacketed stainless steel soup pot that Matt is converting into a small milk bulk tank. The smallest used bulk tank on the market costs at least $3,500 and is much larger than they need. They figure they will have invested between $300 and $400 in their mini bulk tank and may even find themselves a new business opportunity supplying similar equipment for small dairies.

Matt is an excellent welder, so they buy or get free scrap metal, old pipe, metal roof trusses, and other metal to build things they need, such as tractor hitches, hay spears, locking head stanchions for their cows, and mobile pig waterers. Another low-tech innovation they are proud of was immediately fencing their entire property 12 feet inside from the overgrown property line with a double strand of high-tensile wire. This allows them to graze their animals on the entire property and start building soil fertility while providing them the time to properly clear the old property line and make the investment in sturdy, fixed-knot mesh-wire fencing that will keep out coyotes, dogs, feral hogs, large cats, and other predators common in the area.

For the time being they have an affordable, reliable fence line that is easy to maintain, and that is what matters. Another smart, low-cost invention the couple recently made is mobile mineral bars that they move around their fields so the cows can get the minerals they need, and those same minerals will be spread around the fields in the cattle manure.

Every year the Cadmans strive to come up with a budget, then prioritize the investments they want to make that year. They realize they don't have the cash flow to buy everything right away and they need to be patient about scaling things up or improving on their current situation. One large investment, mostly of sweat equity, was made this year when they decided to build an on-farm walk-in freezer to store their meat. This would allow them to raise and slaughter more animals during the growing season and improve storage and inventory efficiency, too.

One challenge the Cadmans met with on the road leading to the decision to build the freezer is that customers are not always able or willing to stock up for the winter during the production part of the year. Ultimately, educating folks that meat chickens can't be grown in winter and that the best time for beef harvest is fall or early summer is their goal. In the meantime, the Cadmans decided to supply their customers' freezer space on-farm.

Through TrueFields LLC Matt and Jerica raise both Milking Shorthorn and

JERICA CADMAN

Getting the essential minerals into the cows and out to pasture

Jersey cows, trying to home in on genetics for animals that can thrive on grass and the intense heat of Texas summers. They hope to begin adding Milking Devon in an effort to develop more moderate-production animals that are very robust and require fewer inputs than the current "race car" dairy cows they have. Under Shady Grove Ranch they raise Red Devon beef cattle, Large Black hogs, Cornish Cross chickens, Broad-Breasted White and Bronze turkeys, and an assortment of brown egg layers. They use non-GMO and soy-free feeds for the pigs, poultry, and dairy cows and don't use any medications unless it is to save the life of an animal. They don't follow any one system of growing but rather incorporate practices that make sense to them and are humane for the animals. They aren't wedded to the heritage breeds they raise—they just want to find hardy animals that do well in their environment and are productive on Texas pasture. Most importantly, they want to produce nutrient-dense, tasty food.

Matt and Jerica's advice for new farmers (they have a lot because they are new farmers themselves)?

- Do your research first, then get out there and try something. Gather your preliminary results, then go research some more.
- Don't reproduce error in bulk. Start out small. Don't gamble too big. You can always expand once your first batch succeeds.

JERICA CADMAN

A walk-in freezer for on-farm meat storage

- Find out what works for you and your animals and your situation. But listen to other farmers and consider their ideas too. There is no one-size-fits-all in farming!
- You can glean good information from even the big guys, so don't write them off just because they are conventional. With their years of experience, chances are they know more about your animals than you do.
- Take stock of your inventory, your budgets, and your available resources. Spend half as much as you think you have available to spend, because things will invariably cost more than you had planned.
- Things will take at least two to five times as long as you expect. Be flexible, and budget generously for error.
- Buy farming books written between 1900 and 1920—all prechemical, preindustrial agriculture.
- Get more training in business management, accounting, tax law, and related topics if you can. Find a good, trustworthy accountant.
- Use your initial profits to build your expansion. Don't scale up until you know the enterprise is profitable. Build those margins into your price from the very beginning.
- Enter business relationships cautiously. Things can go wrong, and if you neglect to spell out the "what ifs" at the beginning, even between the best of friends, it can ruin great friendships and cause you a lot of stress.
- What works for 50 chickens doesn't always work for 500. Start small, but be ready for major changes as you grow.
- Keep your focus pointed toward your future goals. Don't get sidetracked.
- Be good to your customers. Make them feel special for supporting your work; they will reward you. Don't make promises you can't keep, but keep those you do make!
- Keep excellent records! Write down everything, and try to come up with a filing system that is efficient for your operation. One of the biggest time wasters is paper-chasing.
- Don't undercharge! If you don't make money, you will fail. Working full time off-farm to support a farming hobby is not sustainable. Be realistic and fair about your costs (including labor, vehicle use, equipment wear and tear, etc.). If it's too expensive to sell, do one of three things:
- Find a way to make it cheaper, while maintaining quality.
- Find some new customers who will pay what it's worth (and spend a little time educating the rest about the real cost of food).

WHERE TO SHOP?

My favorite place to look for used equipment, vehicles, and building materials is Craigslist, but it is only marginally useful for finding specialized farming equipment. My husband and I use it more to locate building materials or random things to transform into useful farm equipment. For example, we turned an old propane tank, cut in half, into two sturdy, weatherproof pig feeders. By using an aggregator site like SearchTempest, you can look at a wider geography of Craigslist sites and amplify your searching ability. Through SearchTempest we found an old chicken plucker at a private duck-hunting club that was in disrepair and picked it up for a song. After cleaning it, rewiring it, and a few other repairs, we were able to use it successfully on our farm, then turn around and sell it for a handsome profit. Who knows? If you really love tinkering, fixing up old farm equipment just might become a profitable little side business for you!

In addition to Craiglist, try auctions (both live ones and online ones such as eBay), agricultural newspapers, and just driving around farm country looking for old junk that might be lying in people's fields. Also, let every farmer know that you are looking for something—word of mouth can be a powerful form of advertising. When we farmed in California, we were able to find a couple of grain bins, an old egg washer, and old cotton trailers, which we transformed into mobile hen houses—all by driving around and finding them sitting

TANA BUTLER

A converted cotton trailer handily houses hens.

in fields collecting rust. When we were on the lookout for old metal roofing, word of mouth led us to a retired chicken farmer in the same county who wanted some help tearing down his old barns. We lent him a hand dismantling the redundant buildings and in turn were given as much metal roofing as we could haul away.

ALTERNATIVE ENERGY AND FUELS

How you power your equipment, buildings, and vehicles is an important element to discuss. So you want to be more "sustainable," burn less fossil fuels, contribute less carbon to the atmosphere, and be less reliant on the geopolitics of oil? It is amazing to me how progressive so many organic and sustainably minded farmers can be, but when it comes to energy and fuel, they stick to the status quo because of lack of money, time, or (often) creativity. It doesn't take a $100,000 solar system to get you going in the right direction. There are many little things you can do beginning today to help you conserve energy and use more renewable sources. After working as a sustainability manager for a large agricultural firm a couple of years back, I learned a few things about energy use. This is how I suggest you approach it:

1. The most important thing you can do to address energy is to **focus on conservation**. Change your lightbulbs to compact fluorescents or LEDs; install timers or sensors on your lighting systems; take advantage of natural lighting if possible (solar tubes in the ceiling? Change some roof panels from obscure metal to translucent ones?); insulate your coolers better; take obsolete or inefficient equipment off-line; reduce horsepower where you can, and so forth.

   A significant energy hog on farms is refrigeration. Can you reduce the size of your walk-in cooler or freezer or take some off-line through better packing and consolidation? Are your coolers well insulated? You could always add more insulation on the outside with rigid foamboard insulation or even straw bales (watch out that they don't attract rodents). Are your coolers well sealed? If you see ice building up around the seals, that usually means there are leaks. You can also light a match inside by the cooler door to see if it flickers, which may indicate air leaks.

   Keep your compressors running in tip-top shape, too. It might make financial (and environmental) sense to schedule a once-a-year checkup with an HVAC expert to inspect and repair any compressor and seal issues. The energy savings alone will probably pay for his visit. You can also let somebody else deal with the electricity and the infrastructure costs by storing your product in a commercial cold-storage facility instead of on your farm: You will pay rent but probably only a fraction of the overall costs to operate.

2. The second most important area to focus on is **capturing waste electricity and heat**. Are your motors oversized for the task at hand? Are you using high-efficiency motors or low-efficiency ones? Can your water pump be dialed down using a variable speed drive when you need less water? Are your compressors giving off a lot of waste heat? Waste heat can be captured and stored in the form of hot water, which is useful for washing and sterilizing equipment, floors, and so on. Do you have equipment that is plugged in all the time that can actually be unplugged when not in use?
3. **Focus on fuels**. Are certain pieces of equipment (such as water pumps) running on fuel when it might make more sense for them to be running on electricity? Electric pumps can often be connected to renewable energy sources such as solar or wind, whereas diesel-powered pumps cannot. In addition, are you running the most fuel-efficient vehicles you can for the tasks they need to perform? For example, when we used to do deliveries of meat and eggs, if we had small loads we would use a diesel station wagon that got 45 miles per gallon. We would only use the large delivery van for full loads, and even then we got 18 mpg because we chose the most efficient used diesel van we could afford. Are you maintaining your vehicles (tires, oil, etc.) so they get the best mileage they can? Can you use alternative fuels in your vehicles, such as propane, natural gas, biodiesel, or ethanol? Can you grow your own fuels (a recent Organic Valley project showed that it took approximately 10% of a farmer's fields to grow their own biodiesel for all of their tractor use).
4. Once you have done all you can to reduce your need for electricity and fossil fuels, start looking into **renewable sources**. Solar hot water is one of the cheapest and easiest first steps. This goes well with an on-demand water heater to give the water a little bit of extra heat. Other options that can work well on farms and ranches are solar electricity, wind-powered electricity, microhydro systems, geothermal heating/cooling, and biomethane

digesters. Farmers are building their own ethanol stills to process agricultural wastes into homegrown ethanol to run their gas engines. There are many different state and federal programs that help cost-share or provide tax incentives to implement on-farm renewable energy systems. You only need to make the time to look into them.

TAKE-HOME MESSAGES

- Use the right-size equipment. For example, don't use small equipment for big tasks or giant-sized equipment for small tasks. These behaviors are hard on your equipment, hard on your land, and can be dangerous, too.
- Buy used equipment unless you plan to operate it **a lot**.
- Before buying used, do a thorough inspection. If you buy used, you may want to be a decent mechanic, or at least willing to tinker. Otherwise, you could end up spending a lot of dough at the local mechanic's shop.
- Don't use the wrong implement at the wrong time or on the wrong soil.
- Avoid thinking you need every piece of equipment right away; be patient and build your business over time. You want your income to exceed your debt load. Always consider the implications of adding equipment to your cash flow, the ratio of assets to liabilities, and so on.
- Ask yourself if you can borrow or rent. Think hard about how often you will need a piece of equipment. Also, will you need it at the same time as everybody else in your region, making borrowing or renting next to impossible?
- Buy equipment for what you are going to need most of the time, not the specialized equipment you hardly need. For example, if 95% of your tillage can be done using your small 35 horsepower tractor, rent or contract out the other 5% of your tillage needs to someone with a big tractor. How many people do you know who own pickup trucks but only haul something in the back of them once in a blue moon? The rest of the time they could get by and save a ton on fuel by owning a smaller passenger car. Don't let ego creep into your decision: I know some folks would rather be seen on a big tractor or in a big truck rather than an appropriately sized smaller one.

- Get testimonials from others before you invest in a piece of equipment.
- Don't just shop by price. Think long and hard about the longevity and quality of an item. If that cheap polytwine you bought at your local feed store frays and doesn't carry a charge after just one season, you might want to consider buying the expensive mail-order polytwine with a 10-year lifespan. Likewise, it might be foolish to invest in the more expensive piece of equipment with a 20-year life span if you are just experimenting with an enterprise and think you will likely drop that enterprise in a couple of years.

  For example, we put a fair bit of money and sweat equity into building a bunch of bottomless broiler pens, only to decide the next year on a completely different design. Luckily, some of the material we could recycle into the next design, but much of it we couldn't use, nor could we get back the time we had invested, including some hours of paid labor. In hindsight we probably should have only built a few different prototypes, and only one or two of each kind.
- Consider your land tenure situation: Do you have a long-term lease or do you own the land? If so, will it make sense for you to build permanent infrastructure, or could you get by just fine with portable infrastructure? In some cases portable might make more sense, such as on land that frequently floods, or when you want to practice rapid rotational grazing, in which lightweight portable fencing often facilitates better management.
- If you are going to build improvements that add value to the property, you may want to ask your landlord to pay part or all of the costs of those improvements. Don't be shy about asking. Likewise, if you are going to use the Environmental Quality Incentives Program (EQIP) or another grant program to pay for improvements, ask your landlord to pay the up-front costs that will be reimbursed later. Don't make your cash flow suffer for improvements that your landlord will profit from.

CHAPTER 7
SOIL AND WATER MANAGEMENT

THE BANK ACCOUNT that supports your entire farming operation is your soil bank. Despite the foundational importance of the soil, you would be surprised by how little it is attended to by farmers and ranchers around the globe. The way we treat our soil is akin to the way we abuse credit cards. We overspend beyond our means, the debt grows larger, and we can barely afford to make the interest payments, let alone make a dent in the principal. With our soil we continue to take, take, take away the principal nutrients, and our meager supplements of fertilizers do almost nothing to stem that loss. Even though organic farms are supposed to write up and enact a soil management plan, I have seen many of them imitate the same poor practices as nonorganic farmers who aren't required to write a plan. Too many of us act as though topsoil is a renewable resource—it is not.

What are some of the inferior soil management practices that persist to make farms unsustainable, and how do we restore soil health? Much of this section is drawn from the fabulous and thorough publication *Sustainable Soil Management Guide*, published by Appropriate Technology Transfer for Rural Areas (ATTRA), also known as the National Sustainable Agriculture Information Service. Here are what too many of us farmers and ranchers are doing wrong:

1. **Compaction** is rampant. In the quest to get crops in when the soil is too moist, we cause tremendous compaction. In our quest to mechanize everything, we cause compaction. When we use the same implement that reaches to the same depth or flips the soil in the same way, we cause compaction. Unless you are farming or

ranching on "virgin" soil, you might as well assume that you have soil compaction. Now what?

- Rotate your crops, especially between shallow-, medium-, and deep-rooting crops. For example, lettuce and strawberries are shallow rooted, while grains are deep rooted.
- Plant mixed pastures with grasses, legumes, and broad-leaved plants. Their diverse rooting structures will help open up the soil. Also throw in some perennials that develop really deep-rooting structures. This works great in strips between tree crops, too.
- Rent or borrow a deep ripper every few years. You don't have to own it, nor should you come to rely on deep ripping. Yet it can redress serious compaction, particularly deep hardpans.
- Don't plow to the same depth every time, and use different implements when you do plow. For example, relying on only a rototiller that plows to a similar depth every time is a sure cause of compacted hardpan.
- Rotate annuals with perennials, such as vegetable crops with pasture or hay crops. A nice rotation I have seen is two years of vegetable crops followed by three years of pasture. The rooting structure of the pasture plants not only opens up the soil but can also pull up deep nutrients and add significant organic matter.
- Test your soil's moisture content before you bring equipment into the field. Dig up some soil, and squeeze it in your hand. If the soil drips water or holds together like a lump of clay, it is too wet to cultivate.
- Think about how you might be able to extend your season without resorting to working your fields when they are wet. For example, could you start crops as transplants in a greenhouse for a month or two before you set them out in the field? This holds true for livestock as well. Could you have dry shelter for your animals during the rainy seasons so that they aren't trampling all over your fields when the soils are wet and saturated?
- Watch your irrigation so you aren't oversaturating your soils, which could lead to compaction.
- Building soil microbiology and organic matter, correcting pH, and achieving a proper calcium-to-magnesium ratio will all contribute to better-aerated, better-draining soils that are more resistant to compaction.

- Even incessant foot traffic during the rainy season can cause compaction. Enter the fields as infrequently as possible when it is wet.

2. Very few farmers seem to be adding enough **organic matter**. Are you feeding the soil, or just the current crop growing in the ground? As an organic or sustainable farmer, you should set your focus on feeding the soil. Otherwise, all you are doing is following the conventional model of applying externally produced, simple plant nutrients, just enough (or sometimes way more than you need) to get a salable crop. But how can you actually improve soil organic matter over time?

    - Leave behind crop residues, flail-mow them if possible, then reincorporate back into the soil. Or let your livestock out to consume the residues in situ and deposit their rich manure in its place.
    - Rotate with cover crops or green manures. Try to get in at least one season of cover crops every other year, if not every year.
    - Spread biologically active, high-quality compost on your fields once every one to two years. It does not pay to buy cheap, mostly carbonaceous municipal composts. Shop around; ask for the company's most recent compost tests and references of other farmers who have used their compost. Call a reference or two and ask for honest feedback on the compost quality. Applying a good load of compost can be expensive—make sure it is worth your money.
    - Use plant-based mulches, such as leaves, straw, wheat hulls, or wood chips, on the beds or in the walkways. Find out what locally available mulch materials may exist in your region: In some places that might be nut hulls, while other places might have oat or barley hulls. Ask around—you may be surprised what organic waste materials you can find. This organic matter will break down slowly over time and feed your soil microorganisms. However, be careful not to add too much carbonaceous material right near the growing plant, as it can tie up nitrogen and stunt the plant's growth.
    - Rotate livestock into your system, but be careful not to overgraze. Remember that root growth mimics aboveground plant growth. If you want a lot of root growth and decay (and

hence organic matter), don't graze the plants below a healthy level and let them periodically grow much taller.

- Get your soil tested every one to two years. Strive to make your organic matter percentage go up incrementally over time. Make a goal of attaining something like 70% of the native soil organic matter (SOM) level for your region. For example, if the native grasslands where you farm have a SOM level of 8%, your goal level should be 5.6%, which would represent 70% of native levels.
- Organic matter will be useless if your soil is biologically inactive. By adding organic matter and biologically active composts, you will be feeding the microorganisms that live in your soil and helping them proliferate. There may be bacterial and fungal inoculants out there that can help, too, although many have not really been proven to do much other than cost you money. Your money will be better spent on making or buying good compost if you want to simultaneously bring in more beneficial microorganisms and feed your existing populations in the soil. No soil is completely devoid of microbial life, but some may need some outside assistance.

3. **Soil testing** happens too infrequently, or is done improperly, or is not conducted at all. What is $80 or $100 a year in the grand scheme of your farm budget? Why don't you make soil testing a regular thing you do each year? There are so many fascinating parts to a soil test that I can't go into them all. A great book that lays it all out is Neal Kinsey's *Hands-On Agronomy,* which draws heavily from the Albrecht/Reams method. Here are some tricks to making the most of your soil test:

   - Use a reputable testing lab (not a fertilizer company), then stick with them over time. If you bounce around using different labs each time, you will get very different results, making it hard to compare your soil's chemical properties over time. If you are an organic farmer, try to find a lab that has expertise in organics.
   - Don't take soil samples when your soil is too wet, frozen, or parched dry. Take a sample each year at around the same time, ideally two to three months before you plant your crop so you have time to review the results and add some of the recommended amendments.

- When taking your soil samples, follow the protocol described by the testing lab. Usually, this involves randomly selecting 10 spots around your field (I like to throw a bright yellow tennis ball around so I can see where it lands), peel off the top grassy/weedy layer, then dig your clean trowel 7 inches down and pull up a slice. Thoroughly mix those 10 samples together, and send 1 cup of that mixture in a paper sack to the lab immediately. Don't keep the soil in your truck, baking in the sun, for several days before you send it. That will change the results! If you are planting deeper-rooting crops such as trees, you might want to take a separate soil sample at a much greater depth. Above all, follow the sampling directions that the lab gives you.
- Read the lab results thoroughly, especially their recommendations, then have another farmer mentor/friend of yours have a look, too, as a sort of second opinion. Early on in your farming career, this might be where you could bring in a soils consultant to help interpret your results and make recommendations. This is money well spent.

4. Not addressing **major mineral imbalances** (such as calcium-to-magnesium ratios) or pH problems. It always amazes me how commercial farmers, both big and small, neglect the mineral content of their soils. Sure, they probably give some sort of nitrogen-phosphorus-potash (NPK) shot to their crops, but much of that is taken up in the crop. What about the other major and minor minerals that are necessary not only for plant growth but also to support diverse microbial populations and organic matter decomposition? Likewise, I see farmers killing themselves to get a crop out of dense, sticky, saturated soils that probably don't have the right cal:mag ratio or the right pH (which affects hydrogen and oxygen exchange and hence aeration). How can you address the chemical properties of your soil in a way that doesn't break the bank or your back?

    - Start off, before you do anything else, with a soil test.
    - Observe the soils: how they drain or pool water; whether they crack or swell.
    - Feel the soil, and squeeze it through your fingers. Notice the soil texture and color. Dr. Ray Weil, a soil scientist at the University of Maryland, describes how he would make a

quick evaluation of a soil's health in just five minutes: "Look at the surface and see if it is crusted, which tells something about tillage practices used, organic matter, and structure. Pushing a soil probe down to 12 inches, lift out some soil and feel its texture. If a plowpan (hardpan) were present, it would have been felt with the probe. Turn over a shovelful of soil to look for earthworms, and smell for actinomycetes (white fungal 'roots'), which are microorganisms that help compost and stabilize decaying organic matter. Their activity leaves a fresh earthy smell in the soil."

- Add major minerals via compost, cover crops, aged manures, even livestock in rotation. If your livestock are fed the minerals that your soil is missing through their feed or a mineral block, they will help spread those minerals out onto your fields. Other decent sources (especially if locally available and renewable) are kelp/seaweed, fish emulsion, compost or manure teas, and blood, bone, or feather meals. In my opinion, formulated fertilizers (even organically approved ones) are my least favorite source of nutrients: they are expensive, energy intensive, and often too soluble and "fast-acting" as sources of nutrients, and do little or nothing to address soil mineral imbalances. They feed the plant and not the soil, don't provide soil organic matter, don't increase soil microbiology, and can sometimes lead to nutrient pollution via leaching.
- Learn how to do a basic nutrient budget for the three major nutrients of nitrogen, phosphorus, and potassium to calculate your fertility needs. Here is an example of nutrient budgeting for nitrogen:

  Lettuce needs 100 to 120 pounds of nitrogen per acre to produce a decent crop according to UC–Davis Extension bulletins. Let's pick 110 pounds per acre as the target average. Your soil test indicates that 25 pounds of nitrate nitrogen (NO_3) is available in the soil (although nitrate nitrogen is very *dynamic* and can leach out before you even plant your crop, so this number may not be that accurate). Your water test indicates that 50 pounds of nitrate nitrogen (NO_3) will be available over the course of irrigating the lettuce to harvest.

  So if 110 pounds of nitrogen is desired:

  25 pounds nitrogen present in soil, plus
  50 pounds nitrogen present in water
  equals 75 pounds of nitrogen that is available

Which means you will need to add 35 pounds of nitrogen per acre.

If you hadn't done this budget, you might have added 110 pounds of nitrogen per acre when you only needed to add 35 pounds. Knowing this is important for nutrient management and cost containment and is also a water-quality issue that I will discuss further along in this chapter.

5. **Soil erosion** beyond what our soils can naturally regenerate. To me this is truly the most egregious thing we could be doing to our soils. Soil erosion created the downfall of many important ancient civilizations, and it will be the same for our current society. Sometimes, soil erosion is subtle and hard to see; even flat fields experience erosion. It can happen from wind or water. Erosion is not only an enormous problem for our agricultural productivity but also for the health of our waterways and oceans, as they are impaired with excessive sediments. If you do anything to improve your soils, priority #1 should be stopping soil erosion. Keeping your soil where it belongs is the best investment you can make. How can we do this?

    - Minimize bare time. Plant a cover crop during those fallow periods or at least leave all the crop residues on site to protect the soil.
    - Minimize soil disturbance during windy times. If the afternoons are always gusty, can you cultivate in the mornings instead?
    - Don't overirrigate, and quickly fix irrigation leaks. Use drip or micro-sprinklers whenever possible. Irrigation runoff causes a lot of soil erosion.
    - If your land has curves, minimize tillage on the steepest parts. Maybe they are best left for pasture or perennial crops. Align your rows along the contours of the land, which also helps you improve irrigation efficiency. Your local National Resources Conservation Service (NRCS) or Soil and Water Conservation District should be able to help you do row arrangement/contour planting.
    - Plant grass on your field roads and around field edges, especially where runoff may occur.
    - Stand out in the rain. Observe where the rain pools and runs off. What color is the water? Does it run clear, or is it cloudy with sediment?

- Don't get caught with your soil uncovered when you know the big rains are coming. Pay attention to the weather. Don't pull your crop out or plow the week before the rainy season begins.
- Compaction can cause rainfall and irrigation to sheet off instead of penetrating the soil and soaking in. This sheet runoff can cause tremendous erosion of topsoil, so address your compaction issues.
- Overseed cover crops into your cash crops so they can cover the soil once you are done harvesting the cash crop. White clovers or vetches work really well in this type of system in some climates. Experiment to find out what works best where you live.
- Improve soil organic matter and microbial activity, which create the humic acids that "glue" soil particles together to help soil resist erosion.
- Plant windbreaks and hedgerows to minimize soil erosion.
- Last resort: Build a sediment pond to capture runoff and sediments. Once every one to two years, excavate the sediment from this pond and spread it back over your fields, where it came from.

Soil has physical characteristics, chemical characteristics, and biological characteristics. I have tried to cover a bit about all three dimensions of your soil. You must attend to all three to become a successful, sustainable farmer. As a farmer, you can't get away with not modifying the soil properties in some way (unless you are doing soilless plasticulture such as hydroponics or aeroponics). Your job as a sustainable farmer, however, is to improve those soil characteristics over time in a way that is also economically viable. You may not be able to redress all the past soil abuses in a short period of time. You must be patient yet diligent. Don't let soil health slip to the bottom of your meager budget or your overwhelming to-do list.

Keep thinking of your soil as like the bank account that I described at the beginning of this chapter. Are your withdrawals, in the form of crops or livestock sent off the farm, greater than your deposits of fertility? You need to replace those nutrients exported from the farm. Nitrogen-fixing crops are one of the best ways to *add* nutrients. On the other hand, if your deposits are much greater than your withdrawals, you could be causing excess nutrient buildup and likely pollution.

If you are just stockpiling crops or animals and not selling much, this could happen. It doesn't seem like something a commercial farmer would do, but believe me, I have seen it. For example, I have witnessed many farms with unproductive animals that are allowed to continue to reside on the farm. The animals continue to urinate and defecate, of course, and this can lead to manure buildup and a host of other problems (such as disease) if they are not culled.

Although many of the farms we visited employed a myriad of excellent soil conservation and management strategies, the Crown S Ranch in Washington is not only practicing sustainable soil management, they are also monitoring it over time. For example, it used to take them 1.5 acres of pasture to finish a steer; now it takes only half that, or 0.75 acre. This is a direct result of their farm's improved soil, which is now able to support more lush, diverse pasture plants that in turn support the animals on less land.

The family has also excelled at using animal impact and crop rotations to improve the soil and farm productivity at the same time. Read on for how they got started and what changes they have seen over time with their farm management practices, how they are relying less and less on off-farm inputs, and integrating preindustrial farming know-how into their modern-day farm.

Engineered for Success

CROWN S RANCH, WINTHROP, WASHINGTON

• Jennifer Argraves, Louis Sukovaty, and children Geza and Icel •

On 150 acres of land in the Methow Valley of North Central Washington, a small revolution is taking place. In an area that supports mostly monocrop-type production of grains, hay, or cattle, the folks of Crown S Ranch are doing it a little differently. They are layering over six different cropping systems on the same land, including livestock, poultry, grains, legumes, hay crops, and vegetables to produce extraordinary amounts of food per acre and a savings account of fertility for the future.

Owners Louis Sukovaty and Jennifer Argraves returned to Louis's family ranch to raise their own family, after spending 10 years in Seattle as engineers. They each had the rare privilege of growing up on small family farms, raising much of their own food and learning to appreciate it. They wanted the same rich experience for their own children. At first they attempted to run a small engineering firm while living on the ranch, but as their young toddler began eating solid foods, their interest in the food system peaked. Jennifer pored over books and websites looking for information and found that her local food shopping options were coming up short in terms of healthy, natural foods. So they began raising some of their own food and over time yielded to the desire to grow more for other families.

The couple also researched sustainable agriculture and found that many of the university-generated "solutions" were not really addressing the root causes of the problems, such as pest control. It turns out that some of the best practices in sustainable agriculture come from pre–WWII research, before chemicals and nitrogen fertilizers wreaked havoc on our farms.

Jennifer and Louis are taking steps each year to close more cycles and rely less on inputs or stock from off the farm. They now grow 85 to 90% of their feed grains and 100% of their own hay and are reproducing their own animals on the farm—even incubating their own chickens. Their seven-year land rotation has built tremendous fertility; in 10 years of farming together, they have doubled their organic matter percentage.

The couple also built into their rotation what are called "dead-end hosts," which are animals that don't allow specific parasites to reproduce and continue; therefore, the parasite population dwindles and can eventually disappear. This eliminates the need for antiparasitic drugs, which are not only costly but if used continually can actually lead to new drug-resistant parasites. A typical seven-year rotation at Crown S Ranch might look like this: alfalfa, years one and two; cattle, years three and four (rotated around within that time); poultry, year five (also rotated around within that time); and grain years six and seven.

Raising several different species of animals on the same land creates a number of positive returns; Jennifer refers to this as "layering." First, it can produce more food per acre and increase gross profits per acre. It provides a variety of manures that contain different levels of nutrients and trace minerals, providing all the fertility the ranch needs. Different animals have a variety of impacts on the pasture plants and the soil. Some browse higher on the plant, such as sheep and cows that graze at the top of the grass, whereas others clip close to the ground, scratching it up, as chickens do, or even root under the ground, as pigs do. The animals also select for different plant species, so a layered system will provide more even grazing. Multispecies grazing also provides enhanced pest control, such as for parasites and flies.

Jennifer and Louis have put their engineering backgrounds (and minds) to use in a myriad of ways to plan and implement their farming vision. They have mapped the entire ranch on AutoCAD (a computer-aided design tool) so they can keep track of their rotations, field sizes, soil test results, and other characteristics of the land. They also combine traditional animal husbandry with new technology where appropriate. One example is their use of portable electric fencing. While rotational grazing has been common for centuries—often with the use of herdsmen, dogs, and horses to move animals to new locations—at Crown S Ranch they move their animals frequently but keep them contained with electric fencing.

JENNIFER ARGRAVES

Pigs are easily contained on pasture with two hot wires.

The couple also uses the natural behavior of the animals to complement each other, such as running chickens right after their beef cattle. The chickens break down the cowpats, eat the fly larvae, and help recycle nutrients back into the soil. This reduces the number of flies that irritate the cattle and allows the soils to more readily absorb the cowpat nutrients. Thus, the grass will regrow under the cowpats once a chicken has demolished it, whereas it won't grow as well or as quickly if the pat just lies there and dries up like a hard cake.

The couple found an old method of fly control from a University of Nebraska researcher (again pre–WWII) in which their cattle walk through a dark passage that has baffles on the sides with sunlight shining down through them. The face and horn flies on the cattle will instinctively fly up toward the sunlight but get caught in the baffles and eventually die. The dead flies can be composted or even fed to the chickens. Combining the ferocious fly-eating habits of chickens with the passive walk-through fly trap has given Crown S Ranch better fly control than most conventional pesticide fly treatments give. As a bonus the nontoxic flies provide food for their chickens, which is one more enterprise of the farm (free food!).

Another invention of the couple is their solar-powered chicken tractor. A solar panel mounted on the chicken tractor slowly charges a battery, which runs a small motor that moves a set of wheels.

JENNIFER ARGRAVES

Homemade mobile broiler pens moving down the field.

This moves the chicken tractor in small increments, about 4 inches every half an hour, so that the broiler chickens inside can always have access to fresh pasture. This movement increases their consumption of pasture plants and forage like insects and worms, reduces their grain feed consumption, and also creates more healthy conditions for the birds because they are moved off their manure (which fertilizes the soil underneath). Likewise, this invention saves labor for this family that already is extremely busy managing a lot of moving parts.

The couple first began marketing their meat to friends and family. They sold mostly in bulk, such as quarters, but realized only a handful of people have that kind of freezer storage space available. Then they started a meat CSA in the Puget Sound region, the first one of its kind there, with the help of a group of motivated ladies who had recently read Michael Pollan's book, *The Omnivore's Dilemma.* They also added other accounts, such as small stores and a couple of farmers' markets, to the Seattle route to maximize the value of driving that far (around 3.5 to 4 hours one way). More recently, the family opened a farm store on the ranch for them to retail directly to their customers, which is especially important for feeding their more immediate community in Okanagan County (which is considered a "food desert," even though agriculture is a main industry there).

JENNIFER ARGRAVES

The progressive ranchers and their kids

While farms have gotten bigger and bigger across this country, Crown S Ranch is proving that a modest-size family farm that layers systems can be incredibly efficient and profitable, while producing clean air, water, carbon storage, and healthy, nutrient-dense food. It is a model they are happily sharing with others. Each year they invite six or seven interns from around the world to come and learn from each other and gain skills in sustainable agriculture. The couple has developed a special relationship with some of their interns, such as those from the MESA program (MESA stands for "Multicultural Exchange in Sustainable Agriculture"), which provides learning and on-farm training opportunities for young adults from developing nations. Because these folks often have more traditional animal husbandry skills because of where they live, Jennifer and Louis learn as much from them as the interns do from Jennifer and Louis.

The couple is hoping to set up a "sister farm" with one of their former MESA interns from Peru because they think their layered animal system will bring benefits to that community. They try to make a regular habit of going on a farm vacation every couple of years to witness different forms of agriculture. Their latest trip was to Taiwan, where they were amazed by the intensive terraced vegetable production systems there.

Another way the folks of Crown S Ranch are educating the community about sustainable food production is through their recently restored farm cottage, now home to their "Haycation" program. For a reasonable price families can stay in this lovely cottage on the farm, experience a little bit of farm life, eat good food, connect with nature, and go back home a little more willing to support the reemergence of small family farms. Although it is not a profit maker at this point, it does pay for itself. More importantly, it is creating a new breed of consumer advocates willing to invest more heavily into their values.

There are a few things standing in the way of farms such as Crown S Ranch. One is the unlevel playing field in which the giant megafarms get subsidized while the smaller, diversified ones do not. That means that Crown S Ranch meats have a higher price tag because they are paying for all the true costs of production, not externalizing those costs onto others, or relying on governmental aid or subsidies, as is the case with manure pollution control on big farms.

Also, the rising complexity of meat processing regulations has favored the biggest players, which can afford to comply, while shutting out many of the smaller, independent facilities. According to Jennifer's calculation, only seven USDA-inspected abattoirs still exist in Washington State, and they aren't all willing to work with lower-volume producers or ones that want to maintain their brand identity. Crown S Ranch does process some animals in a USDA facility so they can sell retail cuts at markets, to restaurants, and through their own retail space. However, they have designed their meat CSA so their customers own individual animals; thus, the animals can be harvested on the farm and cut and wrapped in a nearby custom facility—a less stressful scenario for both the animals and the farm owners.

Jennifer has made it her mission to fully understand the Washington State and federal regulations. She wishes that the Washington Department of Agriculture would create a state inspection program and opt out of USDA inspection. That would greatly expand the meat processing options for independent producers around the state.

The new USDA meat labeling regulations that come into force in 2012 will also prove costly for Crown S Ranch. They will have to label each cut of meat with nutritional information, which requires

JENNIFER ARGRAVES

Icel inspects the family's beef herd.

not only a costly label but also mislabels their meat, since the USDA-approved label does not distinguish between grain-fed and grassfed animals (the nutritional benefits of grassfed, pasture-based animals is heavily documented). Meanwhile, the big meat packers still don't have to say where their meat came from (country-of-origin labeling has been shelved by the USDA for now) or what it was fed, what it was injected with, or what devastating environmental impacts it created, like the Superfund cleanup sites.

On a positive note Jennifer and Louis did receive a grant a few years back from the USDA Value-Added Producer Grants program. This provided some needed funds to create a professional logo, labels, and signage and to purchase farm record-keeping software.

Jennifer and Louis must be doing something right because their children are excited about the possibility of taking over the farm someday. They both help out quite a bit on the farm and are learning valuable skills and a strong work ethic, regardless of what they choose to do later in life.

Jennifer and Louis's advice for new farmers?

- Volunteer on farms. Farmers need the help—they always have too much work to do and not enough time. Do it as a benefit to the farm—be open to taking on every task, and you will learn from the farmer along the way.
- Everyone has a place in this system—by working on different farms, you can help identify your role. Each person has her own skills that she can add to the new sustainable food system.

WATER MANAGEMENT

We all know, or should know, that clean, fresh water is rare and that geologic sources of it are dwindling. Well levels are dropping as groundwater is depleted, saltwater is intruding into coastal wells, land is even subsiding and sinkholes forming as aquifers are exhausted. Surface water sources are being diverted to inedible lawns, pleasure ponds, and ever more competing users, both rural and urban. What surface waters remain are often polluted, heavy with salts or other concentrates, and teeming with pathogens. Normal water cycles are no longer predictable, with many wet places getting even wetter and arid places more parched. What on earth is a food producer to do?

Water Conservation

First, work on your soils so they have the proper drainage and natural water-holding capacity at the same time. Everything starts with the soil. Farmers have been "drought-proofing" their soils for millennia. Ways to work toward this include augmenting soil organic matter; fostering microbial diversity; ensuring stable humic acids, a proper cal:mag ratio, and appropriate nutrient levels; and using crop rotations with periodic fallows.

Obviously, plant the crops that are most appropriate for your climate and the irrigation water or natural rainfall you have available. Seek out varieties that produce good quality for your markets but are less water intensive. Heirloom or heritage varieties or breeds might be good for that. Encourage your local Extension stations to research more drought-resistant varieties. It always baffles me when the land-grant universities come out with new water-hungry varieties that ignore the water resource constraints of the future, but that rant is a subject for another book altogether!

Invest in the best irrigation system that you can afford. Irrigating right next to the crop is the most cost- and resource-efficient method, and it reduces weed competition because you aren't irrigating a bunch of weed seeds. Consider drip irrigation for vegetables, microsprinklers for tree crops, and low-energy precision application center-pivot systems for field crops. Try to avoid spraying forms of irrigation when it is windy. Constantly check your system, and fix problems when you see them. Make sure your water pressure and volume are the best combination for your system.

Beyond design and installation is use and monitoring. Use a tensiometer (or your hands and eyes) to measure the soil moisture before, during, and after irrigating. Ask your local Extension agents or fellow expert farmers how many centimeters of water per week your crop needs to produce properly. Learn how to apply irrigation to achieve this ideal scenario, then measure the results in your soil and by the appearance of the plant. I have met so many farmers who irrigate whenever their plants look thirsty (which is too late) or by some other random indicator, and they typically have no idea how much water they are applying to their crop. Overirrigating can cause many problems, from reducing nutrient exchange in your soil to causing excessive leaf growth in your crops and reduced fruit set. Overly watery crops can attract more pests, and when harvested they can rot faster.

In contrast some crops can even be grown with little supplemental irrigation (sometimes referred to as "dry farming"), relying on natural soil moisture and rainfall to produce. Grains are commonly dry-farmed, but even some vegetable crops, tree nuts, and tree fruits can be dry-farmed once they get established, as long as the evapotranspiration rate is not too high. My favorite dry-farmed crop is tomatoes—so dense and sweet I can hardly go back to eating irrigated tomatoes ever again.

If you raise animals, there are ways to save water, too. Don't overfill your water troughs so they are always spilling over. This will also prevent some of the mud and pugging that happens around water troughs. If you use a float valve, position it a little lower so the trough doesn't fill to the very top. Automatic waterers can be a great way to save water and keep the quality of the water better for your animals. Make sure you check them frequently for leaks so they don't waste a bunch of water and create damp, disease-causing conditions for your animals. Provide shade over your water to decrease evapotranspiration loss and keep the water temperature cooler for your animals. This applies to open water sources, too, such as ponds. Plant some trees around the pond to provide shade and reduce evapotranspiration loss.

Water Quality

As an organic or sustainable farmer, you are probably using much less pesticide and fertilizer than a conventional, chemical farmer. This fact bodes well for water quality. However, you can still cause deleterious impacts on water quality, and you should remain hypervigilant. Even organic pesticides shouldn't get into our waterways, nor should nutrients coming from your compost, aged manures, fish emulsions, and other organic fertilizers. Try not to use any of these materials within 100 feet of waterways. Likewise, implement practices to reduce rain and irrigation runoff, which could carry pollutants into nearby waterways. Excess nutrients can also contaminate groundwater sources through leaching. Avoid applying fertilizer or spreading manure on your fields before a big rain event, and don't overirrigate after applying either.

Even with organic amendments you can still calculate the rough NPK amounts—make sure they correspond to what the crop needs. Don't forget to factor in the residual soil NPK and what might be in the irrigation water. In many places nitrates are common in the well water. Irrigating over the crop season may give you up to 100 pounds of free nitrogen! Make sure you factor that into your calculations. Excess nutrients the plant doesn't need, particularly soluble ones that are more likely to leach out, contribute to water pollution.

When your crops need nutrients such as nitrogen beyond what the soil and irrigation water may contain (based on your nutrient budget), additional pounds of it can be applied through what is known as "fertigation," in which you apply fertilizers directly via the irrigation system. A lettuce farmer who probably uses drip lines for irrigation could pump a soluble organic fertilizer or compost tea into the drip lines at a rate that supplies those additional pounds of nitrogen. However, I don't love purchasing formulated fertilizers, which tend to be energy intensive and come from some questionable sources, so another option for future plantings is to plant a nitrogen-fixing crop such as beans or cover crops such as vetch before your lettuce. Other crops that provide a lot of nitrogen are brassicas such as broccoli or mustard, if you till the residues back into the soil.

Water Sources

I said in chapter 3 that understanding your water source(s) was fundamental in your decision of what land to farm. These are some of the key water source factors to ponder:

- **Groundwater or surface water?** Groundwater tends to be of better quality and freer of pathogens than surface water, but not always. Using groundwater irrigation may result in higher electricity and pumping costs, and you will probably be responsible for the well maintenance and repair.
- **Metered or not?** Will you have to pay by the inch or acre-foot, or is the water free except for the cost of pumping it? One benefit of having a meter is that you can more accurately calculate how much water you are applying to the crop. That is hard to know when you don't have a flow meter.
- What is the **volume of water and at what pressure** is it coming out? Will that volume and pressure work for the irrigation system you have in mind? If not, can you adjust it through some sort of variable-speed device on the pump or by sending excess volume into storage tanks to be used at a later time? One farm that I managed had a very high-volume well pump that pumped more volume than our drip irrigation systems needed. With the help of the USDA EQIP program we installed some large water tanks on the top of the hill that we pumped the excess water volume into, then gravity-fed off those tanks when we wanted to irrigate smaller blocks, such as our native plant hedgerows.
- **Water rights**? I could not even begin to cover the myriad of laws and complexity of water rights across the country. I would just say that you should understand what the local water rights laws

are in the area where you want to farm and what rights, if any, come with the specific parcel of land. For example, are you allowed to capture rainfall through roof-water catchment systems or rainfall detention ponds? Can you develop the springs you have on your property or divert some of the creek water? If you have water rights to the nearby creek or river, how much do you get, when, and what is the quality of that water? How much decision-making power will you have in how those surface water rights are managed?
- Is your water conveyed to you through open-top ditches or piped water delivered by some sort of **water district**? How much does it cost to use that water? How is the district managed? Is the water tested, and can you get those results regularly?

TAKE-HOME MESSAGES

- Feed the soil, not the plant!
- Get your soils tested, and update the tests every two to three years.
- Give back more nutrients than you take from the soil (fill that savings account!).
- Rest and rotate your fields.
- Control disease and parasites beginning with healthy soils and how you irrigate.
- It takes nature approximately 1,000 years to build an inch of soil. Don't be nonchalant about erosion. Don't let it happen. If you can see visible signs of erosion, it is already too late.
- Consider ways to reduce the physical tillage of the soil, especially tillage that inverts the soil layers or causes a plowpan (compaction layer).
- Get a soil map and understand the physical, biological, and chemical properties of your soil. Be a soil detective. Watch for cues from the vegetation that grows on or nearby your fields.
- Chemical fertilizers cause salt buildup, increase disease, and leach into waterways or aquifers and are energy intensive to produce, to name a few of their problems. Organic fertilizers can be expensive, energy intensive, or shipped from far away and can also cause nutrient imbalances and can leach, too. Think of ways to create homegrown in situ fertility (nitrogen-fixing cover crops, compost, aged manures, rotations with animals, compost teas, pasture rotations, stubble incorporation, mulches, etc.).

- Don't let your source of fertility become somebody else's source of pollution. This is especially true for livestock and poultry. Your manure pile aging 20 feet away from the creek could be polluting the creek with nitrogen or phosphorus. Your rotational grazing of animals through a watershed could be introducing fecal coliform or *E. coli* into otherwise pristine creeks. You may still be having negative environmental impacts even while practicing so-called sustainable agriculture. Keep an eye on the *whole* system.
- Have a good grasp on your farm's water sources, their quantity and quality throughout the year. Don't find out later that the pond you plan to irrigate from has high levels of fecal coliform bacteria or that the stream you will pump from goes dry for three months each summer.
- Have a backup water supply for emergencies and droughts (e.g., water tank, pond, developed spring, nearby creek, second well, city connection, etc.).
- Protect the quality of your water source. Build a vegetated buffer around your well that you don't till or put animals on; protect riparian buffers along waterways; promote infiltration and aquifer recharge through swales, keyline ponds, constructed wetlands, limited impervious surfaces, and other water quality practices. Keep adequate distance between petroleum products, chemicals, manures, and other potential pollutants and your sources of irrigation.
- Know how you are going to irrigate before you plant anything. If you live in rain-fed agricultural areas, have a backup plan for irrigating high-value crops or animals during times of seasonal drought.
- Invest in high-quality irrigation supplies and parts. Stay on top of maintenance and repair leaks quickly. Irrigation leaks not only ruin crops, promote disease and weeds, create muck and compaction, and leach out valuable nutrients, but they also create pollution problems elsewhere.
- Shop around for irrigation supplies—prices can vary widely. Make friends with the dealer—he can often do design and installation for free or a reduced price if you buy the parts from him.

CHAPTER 8
HARVEST AND PROCESSING

So YOU HAVE ACQUIRED the land, identified your potential markets, figured out how to finance your startup or expansion, developed goals and a written plan, figured out what kind of equipment and infrastructure you need, and understand how you are going to take care of the natural resources that you depend on. When determining which blend of enterprises to pursue, you need to think deeply about how you are going to harvest the crops or animals, process them if need be, and move them into the marketplace. And you have to do all that in a way that is economical, is environmentally friendly, maintains quality and food safety, and doesn't drive you absolutely crazy. Since there are countless books on crop and animal production, I want to focus on one aspect of production that doesn't get mentioned enough, and that is **preparing your goods for market**.

Each crop and animal has specific harvesting and postharvest handling requirements; that subject could be an entire book unto itself. Here I will just cast a broad net—a mile wide and an inch deep—about harvesting and processing different crops, with ideas and questions to ponder during the planning and execution phases, as well as some creative solutions to harvest and processing challenges.

FRESH-PRODUCE HARVEST AND HANDLING

If you're going to grow fresh produce, you should first develop a detailed cropping plan. You can elect to do this in a computer spreadsheet, using an online program such as AgSquared, or on a large piece of paper that is more visual (the last one is my preferred route, but I am a visual learner). Whichever method you choose, make sure your cropping plan includes the previous rotations, where you are planting each crop and when, seeding rates, seed source, and days to harvest.

If you are growing for fresh-market sales, you probably will want to plan for staggered plantings or grow different varieties so you can have more of a crop for a longer period of time. The benefits of extending your cropping season include improving cash flows, increasing gross profits, providing longer-term employment (which tends to keep good people around), and keeping your presence and brand visible in the marketplace. Hoch Orchard & Gardens, profiled in chapter 11, does just this with fruit trees, planting a diversity of apple varieties so that they can sell fresh apples from August through December. This also has the added benefit of spacing out your labor hours so that instead of having a crazy two-week period of harvesting day and night, you can harvest less total volume over a more spread-out period of time.

Once you have a cropping plan, you should also develop a harvest plan. Here is an example below of what you might want to include in a harvest plan, with details about the crop, variety, location, harvest dates, expected yields, and notes about specific harvest and post-harvest handling steps that are key for you or employees to remember. If you raise animals, you can create a similar plan for their processing.

Some great resources to learn more about standard harvest and postharvest handling specifications, including typical grading and pack sizes, can be found at:

Penn State: http://agmarketing.extension.psu.edu/wholesale/prodpkgguide.html

UC–Davis: http://postharvest.ucdavis.edu/producefacts/

Keep in mind that if you are going to be direct marketing via farmers' markets, a roadside stand, or similar venues you don't have to pay as much attention to traditional grading and pack sizes that are described in the websites above. You can harvest an assortment of sizes and allow your customers to choose which ones they want or make bunches themselves, potentially reducing your labor expenses. Sometimes prebagging or prebunching crops will help you sell

SAMPLE HARVEST PLAN

CROP & VARIETY	LOCATION	DATES OF HARVEST (PREDICTED)	EXPECTED YIELDS	PROCESS
Lettuce (*Little Gem*)	Field 1: Rows 2–8	June 15, June 30, July 15, July 30, Aug 15, Aug 30	200 heads each harvest	Harvest before 10 a.m.; pack 24 to a box with top liner; store at 32°F
Radishes (*French Breakfast*)	Field 1: Row 1	June 1, June 8, June 15, June 22	50 bunches each harvest	Harvest early in the day into blue totes; bring to cooler and wash off tops and roots; make into bunches with 6–7 radishes each; pack 24 bunches to a box; store at 32°F
Sweet Peppers (*Corno di Toro*)	Field 2: Row 4	August 1–30	100 pounds over August	Harvest some yellow and some red but pack separately, early in the day, into blue totes; bring to cooler and wash gently; then pack according to size (small: 90, med: 75–85, large: 65); store at 45°F (nightshade cooler)
Broccoli (*NutriBud*)	Field 2: Rows 1–2	1st planting: June 15–30, 2nd planting: Oct 1–15	150 lbs per planting	Harvest when heads are full, early in the day; bunch 2–3 heads together; field-pack 18 bunches to a box; return to cooler and add ice to the top with liner on top; store at 32°F

more of them; that is certainly the conventional wisdom in a lot of mainstream grocery stores. On the flip side some customers like the choice of picking out their own individual fruits or vegetables and might become loyal customers if you allow them to do that. You can also reduce solid waste if you don't put everything into plastic bags or clamshell containers with rubber bands or twist ties around them. If environmental stewardship is one of your production values, then you will demonstrate that commitment through the way you package and market your products.

Try to keep lower-quality grades out of the mix: perhaps separate them into a different box, mark with an obvious sign and lower price, and let people decide if they want to buy the lower-quality product. If you are going to sell to intermediary wholesale markets, such as to restaurants, to caterers, or directly to stores, ask them what kind of bunch size, pack size, and grades they would be interested in and willing to buy from you. A restaurant that is just going to process your produce into something else might care less about the different sizes of the vegetable or having things prebunched. Likewise, they may be willing to take blemished product if they are just going to cook it anyway. A food processor such as a cidery might take blemished apples, even ones that have a few worms in them. To be profitable you absolutely should look into markets for those lower-grade products, but don't make low quality your brand.

Additionally, don't be lazy about harvesting, such as sending a mixed box of zucchini to a restaurant with tiny, fancy-grade zukes mixed in with monster, baseball-bat-size ones. That is a great way to destroy the trust you have built with that establishment and turn them off to future sales. I remember once helping a new farmer prepare for a restaurant order of several boxes of fancy-grade zucchini. Even though I showed her the size I wanted her to harvest (per chef instructions), when I showed up to inspect the boxes they were full of tiny zucchinis the size of a Sharpie pen, medium-size ones you normally see at the grocery store, and monster ones that were hollow and mealy inside. She blamed the problem on her workers, who, in her words, "just didn't understand my directions." That excuse does not sit well with me, nor with any produce buyer out there. As the farmer and the boss, you must oversee the proper harvest and packing of your produce.

If you are going to "process" your produce in any way, such as cutting up spinach to bag it, cutting carrots into sticks, turning cabbage heads into a coleslaw mix, or even further processing, you will be opening up yourself to a whole other level of regulatory complexity. Even making a pretty salad mix will probably require some sort of approval from your state department of agriculture or your county department of health and a food safety plan. I remember a small farm I apprenticed on back in the late 1990s, where we used to hand-harvest leaves of various lettuces, spinaches, mustards, and edible flowers into a large plastic tote. Then we took the full tote back to an old claw-foot bathtub that was sitting outside under a giant shade tree. We rinsed out the tub of any dust, leaves, or insects that had fallen into it, then filled it up with spring water.

We put a large metal screen in the tub, poured the greens onto the screen, and plunged the screen down into the tub to wash the greens. We would let the greens sit in the water for about 10 minutes, then pull up the screen, turn it sideways over the top of the tub, and let the greens dry off for another 10 minutes or so. Once the greens were dry enough, we dumped them into another large, clean plastic tote with a locking lid and took these salad totes directly to nearby restaurants the same day. The chefs would give us the previous week's tote for us to take home and clean. While this method was imprecise and would probably raise the eyebrows of some hypersensitive food safety inspectors, the restaurants loved it. They said our salad mix would stay more vibrant and fresh for longer and did not get soggy like some of the commercial salad mixes they were used to. We attributed this to the thorough drying of the salad on the screens and using plastic totes that gave the salad mix a little more space to "breathe."

I think the only thing I would change now would be to have the tub located inside a clean outbuilding so that dust and leaves would not get into it, and I would have the spring water tested annually. Because the water was coming out of a pristine, forested hillside, we assumed it was clean but probably should have tested it! On the topic of salad mixes, I know some small-scale farmers who have built large salad spinners out of old top-loading washing machines. There is also an expensive plastic 5-gallon salad spinner on the market, complete with electric motor. Honestly, you will probably outgrow that size so quickly that spending the $200 or $300 to buy it is probably not worthwhile. I love repurposing things anyway, so the washing-machine-turned-salad-spinner is one of my favorite ideas!

The other things to keep in mind when harvesting crops for market is whether or not to trim off the roots or leaves. Some farmers believe leaving the leaves on (such as keeping carrot tops on) will keep the crop fresher longer. Other farmers will harvest a crop with the full root system to prevent premature wilting, such as with basil or garbanzo beans. Customers can often keep the plants fresh and even growing to some degree if they submerge the root system into a glass of water. Obviously, if you want to harvest multiple times off the same plant, you would not want to pull up the entire plant. Whereas one farmer might harvest the main head off a broccoli plant, then till it under, another farmer might continue to harvest the smaller florets that the plant continues to produce over the following weeks. If you have a market for smaller florets, this might make sense and will add more value to the individual crop if you continue to let the plants resprout.

PRODUCE COLD STORAGE AND PACKING

If you currently don't have a real cooler to store your produce, at least try to harvest during the coolest times of the day and immediately take that product to market before it has a chance to heat up. I know a lot of farmers who deliver in the middle of the night or early in the morning to avoid the heat of the day during transportation as well. As you scale up and prove your concept is viable, perhaps you can invest in a walk-in cooler. Plenty of farmers have figured out how to do this on the cheap, by renovating a used cooler, insulating an existing building, or building one from scratch using used shipping containers or even straw bales.

Whatever you build, try to make sure it is well insulated and energy efficient, drains properly so that water does not pool up inside it, provides some humidity, can be washed easily, and is sealed from rodents and insects if possible. A diversified produce farm I used to work for had two room coolers—one for the 32°F fruits and vegetables and the other for the 45°F vegetables (mostly the nightshades: potatoes, tomatoes, peppers, etc.), which provided ideal conditions for the different crops.

If a ready supply of electricity is available, mechanical refrigeration systems provide the most reliable cooling source, but they can be costly to run. Methods include room cooling, forced-air cooling, and evaporative cooling. A variety of portable forced-air coolers have been designed for use by small-scale growers. However, a range of simple methods exists for cooling produce where electricity is unavailable or too expensive. Some examples of alternative systems include night air ventilation, radiant cooling, evaporative cooling, the use of ice and underground (root cellars or caves) or high-altitude storage. Ice can be used either directly as package ice, to cool water for use in a hydrocooler, or as an ice bank for a small forced-air or room-cooling system. If you are using ice directly in contact with the crop, make sure you use a potable source of water to make the ice.

If you are going to be using a lot of ice, consider investing in a decent ice machine. It can be purchased used, but make sure it is mechanically sound. Buying ice may seem like an insignificant expense, but it can really add up over the year. I met a farmer during our travels who is spending $6,000 a year on ice alone—the farmer could have bought two decent ice machines for that price!

In addition, shade should be provided over harvested produce, for packing areas, for buildings used for cooling and storage, and for transport vehicles. Using shade wherever possible will help to reduce the temperatures of incoming produce and will reduce subsequent cooling costs (it will also make your employees happier). Trees are a fine source of shade and can reduce ambient temperatures around packinghouses and storage areas. Light colors on buildings will reflect light (and heat) and reduce heat load. Sometimes spending money will save money, as when purchasing energy-efficient lighting equipment, which doesn't give off any heat and uses less electricity.

Another aspect to consider when handling fruits and vegetables is the relative humidity of the storage environment. Loss of water from produce contributes to a loss of quality, as visual changes such as wilting or shriveling and textural changes can take place. If using mechanical refrigeration for cooling, the larger the area of the refrigerator coils, the higher the relative humidity in the coldroom. The best method of increasing relative humidity is to reduce temperature. Another method is to add moisture to the air around the produce in the form of mists, sprays, or, as a last resort, by wetting the coldroom floor—just make sure you don't make it too soggy in the room, which can cause bacteria and fungi to spread.

Another way is to use vapor barriers such as wax-coated boxes, polyethylene liners in boxes, or a variety of inexpensive and recyclable packaging materials. Just remember that many of these things are not reusable, recyclable, or compostable and can contribute to your solid waste problem or that of your supply chain. Increasingly, supply chains are asking for less waste. An alternative might be packing into reusable plastic totes that can be sterilized regularly. I know of one organic farmer in California, T & D Willey Farms, that packs into paper-lined wooden crates that are not only reusable but also help promote their brand and give them an old-timey look when stocked in the produce section of grocers they sell to. Remember that your packaging is an extension of your values and your brand. Don't profess to be about environmental sustainability, then pack into highly toxic plastic or nonrecyclable containers. Customers will see right through that as "greenwashing."

GRAIN/BEAN/FORAGE HARVEST AND HANDLING

I have little experience with growing grains, beans, or forages for harvest, but my family and I did spend some time with several of these kinds of producers during our travels (Massa Organics, Butterworks Farm, Crown S Ranch, and Bluebird Grain Farms; see their individual profiles). I also spent six months working for the largest organic rice producer in the country, so I know a little about rice. My family raised fodder crops and forages for livestock, but we allowed the animals to self-harvest them. Apart from some periodic hay baling that we contracted out only when our pastures got ahead of us, our animals did all of the forage harvesting. We eventually got to the point where we used the pigs to seed the crops as well, an innovative idea I will explain a little later in this section. To learn even more about grains, I'm also delving into the literature a bit right now, reading Gene Logsdon's revised classic *Small-Scale Grain Raising* and learning how to bake my own breads from different varieties of grains.

Some of the things I have come to understand about producing grains is that they can be incredibly tricky to master, especially for a new farmer. There are so many issues to consider, from tillage and seeding to fertility and weed control to disease. Then you have the timing of the harvest, finding the right equipment to do it, and much more. At least you can store them for quite some time, often waiting until the price goes up to turn around and sell. You can't really do that with produce. Another challenge is that new-generation farmers often won't have the capital to purchase the expensive equipment needed for seeding and harvesting, let alone the processing of grains. But grains can be done on a small scale while you are in the experimentation phase and even harvested and processed basically by hand. That is how Gene Logsdon describes it in his book, and what people around the world have been doing for centuries. Bluebird Grain Farms, profiled in this chapter's narrative profile, started out with just a $1,000 used combine and an old tractor.

Look and ask around to see which grain varieties other farmers are producing in your region, but don't let that stop you from experimenting. Where I live now, soft white wheat reigns supreme, but other types of wheat can be grown here, too, as well as other grains, legumes, oilseeds, and forages. There are people here raising hard red spring wheat, barley, alfalfa, and black oil sunflowers, to name a few. In the drier, high-altitude plateaus of the eastern Columbia Gorge where I live, I think the Andean staple crop quinoa would grow quite well.

Grains, beans, and forages can be an incredible benefit to your crop rotation, even if they aren't going to be a main enterprise in your business model. They can help control weeds, add organic matter and nutrients, control erosion, improve wildlife habitat, allow you to produce your own animal feeds, and provide many other positive returns. As you build your production skills over time and improve the quality of the grains, you may even be able to sell them as human food.

This is precisely what happened with the Lazors of Butterworks Farm, profiled in chapter 12. After years of producing and milling grains for their dairy cows, they got to the point where they felt confident producing human-grade grains. The Lazors now sell a wide variety of edible grains and beans, including whole-wheat flour, coarse-ground cornmeal, and dried black beans, to name a few. I actually spent a few uncomfortably hot days weeding one of their bean patches that was full of buckwheat that had self-seeded. Because the Lazors are always experimenting and perfecting their rotations, not everything always works according to plan—but this is where their innovations are born.

Be careful about what you rotate into your fields, though; you don't want a grain crop that goes to seed only to become a future weed problem for you. Luckily, buckwheat is an annual that is easy to eradicate, and its flowers attract a suite of beneficial insects and pollinators. Likewise, if your fields are close to other farmland, be conscientious about which grains you grow, so that they don't become weed problems for your neighboring farmers. It's always a good idea to keep people happy on both sides of the fence!

For instance, we were considering raising amaranth as chicken feed, letting it go to seed so the hens could self-harvest the protein-rich seeds and eat the leaves, too. A vegetable farmer on adjoining acreage made it clear to us that he was not interested in seeing more amaranth weeds in his field; he already had enough pigweed, another type of amaranth. We took his concerns into consideration and decided to try pearl millet instead, a crop that was not going to become weedy. The chickens loved the millet, and it did quite well during the hot, dry California summers, even regrowing after a couple of hay cuttings. So we had a terrific crop trial and kept our neighbor happy at the same time!

This brings me to another point: growing grains, beans, and forages for animal feed. Many farmers we have met on our travels wax poetic about growing all their animal feeds as the key to solving their feed-cost problems. Indeed, it can work with the right set of resources and be profitable. But getting into feed production also can be very costly. For example, one cattle ranch we visited had hundreds of thousands of dollars of farm equipment for baling their own forages and chopping their own silage. However, their forage quality was really poor due to low soil organic matter, soil compaction, prolonged drought, and overgrazing. They would have to add quite a bit of chopped alfalfa and molasses to the hay to even get their cattle to eat it. We suggested they buy in cheaper and higher-quality forages from an area with plentiful rainfall and use their giant stack of moldy, low-quality hay to spread on the fields as a source of organic matter and seed.

In addition, we thought that they could eventually do away with haymaking altogether through better rotational grazing, and suggested they sell off all their big hay- and silage-making equipment. Not only would that free up some capital, but it would also keep them from perpetuating the same mistakes. By removing organic matter from their poor-quality pastures via baling, they were actually exacerbating the pasture's poor condition. It was a downward sinking spiral, but a solvable one.

If you have the right equipment (or can beg, borrow, or rent it), as well as the right kind of soils, land costs, and so on, it might make sense to grow some select feed grains for your animals. As I mentioned above, we grew pearl millet for our chickens, along with a sorghum/sudangrass hybrid, and we seeded a diverse mix of perennials and annuals into our pastures that the chickens would find palatable at different times of the year. Although it took some time, we did finally get some nice perennial clover established in our pastures, which the chickens went crazy for. We moved the chickens about so they could self-harvest the forages as they pleased without overgrazing them. For pigs we tried a variety of fodder crops, including rape, mustard, and fava beans. Our clay soils wouldn't allow us to experiment with all the possibilities out there, such as corn or fodder beets, but we have seen other farmers plant those successfully for their hogs.

Misty Brook Farm in Massachusetts tried something we always dreamed of doing. They spread corn seed on a pasture that was too rocky to till and let their hogs in to eat it, dig up the soil, and plant some of the corn seed in the ground through their movements. Then the farmers moved the pigs out after a few days to a different pasture, separated from the first area by an electric fence. The pig-seeded corn grew a beautiful, thick stand, and once it was mature, the farmers let the pigs back into the field to self-harvest the corn. Pigs will actually knock over the cornstalks to get the ears, along with eating the leaves and stripping them down to just stubble. At the same time they are turning a lot of organic matter back into the soil, defecating and fertilizing the soil, and preparing the field for the next crop. In this example, the farmers are saving money by not having to till the soil, saving gas, improving their soil, and allowing the pigs to have some fun while feeding them well at the same time. You should see the hogs go wild when they are unleashed into the corn patch.

A big deterrent to getting into grain production is the typically low commodity prices that most grains command. Massa Organics, profiled in chapter 9, figured out how to grow high-quality organic brown rice and market it for a fair price directly to food service companies and to consumers, earning much more than commodity pricing. The Massa family branded a product that is normally sold as anonymous, blended together with thousands of other acres of rice, and they get a price that is actually profitable. Bluebird Grain Farms of Washington, described below, is doing the same thing with their ancient varieties of wheat, selling directly to consumers.

If you add sustainability attributes to your grains, such as with certified organic or some sort of wildlife-friendly certification (Bluebird has Salmon-Safe certification for example because their neighboring waterways are salmon-bearing), process and store it well, sell it fresh, and develop a well-loved brand, you can command much better pricing than standard commodities—at least that's the Bluebird model of adding value.

Grains, beans, oilseeds, and quality forages can be viable enterprises. Imagine a fresh, cold-pressed, organic sunflower or sesame oil that you sell people in refillable bottles. Or a dry-bean soup blend complete with the spices that you grew yourself, ready to go into the slow cooker. Additionally, some farms grow nutrient-dense organic forages and sell premium hay to livestock owners that are willing to pay for good feed. Horse owners are particularly drawn to high-quality hay for their expensive pets. Organic dairy farms that don't grow enough of their own feed might also be a good customer for quality organic forages. Lower-quality organic hay can be sold as bedding, or you can keep it yourself to bed down your own animals over the winter.

LIVESTOCK AND POULTRY PROCESSING

Processing animal products can take an immense amount of logistics, specialized equipment, animal handling skills, and infrastructure: It's no wonder so many producers sell off their animals to somebody else to do the processing. Dairy animals have to be kept in peak health and milked daily during their season, and the milk has to be dealt with safely and quickly. When fresh-market fluid milk is not really an option, some farmers turn it into dairy products that can store better, such as ice cream or cheeses. A woman I met recently who lived about eight hours from any major population centers decided she would turn her yak milk into butter and cheese, so that she only had to make a trip to the city once a month. Your market will strongly influence the way you go about processing your animal products.

Similarly, eggs need to be collected daily, washed, and stored in a cool place, preferably around 45°F. When we produced eggs we didn't have a full-fledged walk-in cooler, but we did put a swamp cooler in a window of an insulated garage, turning it into an effective cooler. During the winter we just put the eggs in large chest coolers and left the lids open during the middle of the night to absorb the cold night air.

Essentially, for animals that you will process into meat, you have two processing routes you can follow. You can get your animals slaughtered and butchered under USDA inspection, which makes it ready for retail sales into just about any market. Or you can have the animals custom-slaughtered for private sales to individuals, where the person essentially owns the animal beforehand and contracts with you to have it slaughtered. Some states (currently about half of the 50 states) offer a state inspection program, too, which allows you to retail that meat within the same state where it is processed. But increasingly, states are dropping their meat inspection programs (unfortunately), so I don't know how reliable it is to plan your meat business around state inspection. Certainly, use it if you can.

I have written extensively on the meat-processing bottleneck for my blog: Most states simply don't have enough independent facilities to process for the increasing number of ranchers who want to market their own meat, which stifles the proliferation of sustainable, pasture-based meat production. Some producers are taking it into their own hands by building their own on-farm facilities, but that option is only for those with enough volume and capital and a strong enough resolve to wade through the USDA inspection process. I also prefer in-building infrastructure that will serve many users; not only is it a more sustainable use of building materials as well as land, but it also can lower the per-animal costs, too.

A great model that is working for 65 animal producers in the San Juan Islands of Washington is called the Island Grown Farmers Cooperative. Cooperative ownership not only allowed them to raise the capital more easily, but it also ensured the supply of animals that it takes to keep a USDA mobile slaughter trailer and butcher shop open year-round and offering affordable processing.

Whatever meat-processing facility you think you are going to use, ask a lot of questions, get detailed price sheets, and ask for references. The best way to know if an abattoir does a good job is to talk to other farmers who have used them. It is also important to know which parts of the animal you will get back, which is key for you to maximize your per-animal profitability. Above all, remember that the slaughterhouse and butcher are contractors; they work for you to get paid. Just as you would not pay your building contractor if she did a shoddy job on your house, you should expect your meat processor to hold up his end of the contract to get paid. Give him detailed cutting instructions and remind him to follow them.

Of course, you can use another facility if you are unhappy with one facility's quality, but it's never that easy to make the switch. If you have to have new labels made for each abattoir that you use, those costs can add up, not to mention the wait time it takes to have those labels approved by the USDA Food Safety Inspection Service. In addition, ask if you can take a tour of the facility, and use your eyes, ears, and nose to help you make your evaluation. One custom butcher shop we toured had a really putrid smell inside their carcass cooler, so we went a different direction. I am glad we toured the facility before we dropped thousands of dollars on processing and might have ended up with off-flavored meat.

Learn from master butchers how to maximize your carcass yields and what kind of cuts customers are looking for. Understand that if you make everything boneless, you won't get paid for the weight of all that bone. However, you could sell the bones in a different way—perhaps to people who want to make their own bone broth or others looking for some dog bones. We kept the bones in most of our meat cuts, not only to get paid for that weight, but also because the bones actually enhanced the cooking and eating qualities of the meat. We did offer a boneless roast or two, such as the Boston butt pork roast or the occasional boneless loin roast.

Keep in mind the time of the year that you will be selling the meat—is it summer barbecue season or is it winter roast season? Also, size your meat cuts according to the kind of customers you are expecting. Don't leave all your hams intact if most of your customers are single or have small families. You will just sit on those 8-pound roasts forever. Package your steaks and chops no more than two to a package, and in some cases pack just one. We had many single people who appreciated that we packed some of our chops one to a package. Keep in mind that this may make your packaging costs go up, so factor that into your price.

We stored and sold our meat frozen, which provided a lot more flexibility to us and allowed us to travel to the slaughterhouse only once a quarter or so. If your market demands fresh meat (as do many restaurants), you will have to weigh the increased logistical costs of taking animals more frequently to slaughter to see if it is worth it for you. I know one farmer who takes a couple of pigs to the processor each week, an eight-hour round trip, so that he can sell fresh pork to restaurants and retailers. I am not sure how anyone could justify taking a full day out of his workweek, plus all the wear and tear on the vehicle, for one or two pigs. I suppose if the meat price is high enough, it could be worth it. I personally would find an employee to do this for me; my back and psyche couldn't handle that much driving.

For custom-processed animal sales to be legal, you have to presell the animal to somebody (sometimes to several folks, each purchasing a quarter or half). The new owner then contracts your business to take that animal to a custom slaughterhouse or have a mobile slaughter trailer come out and slaughter on the farm. The customer-owner then arranges the butchery herself and pays for it, too. Some farms handle all the logistics and have one price that covers all the steps; others will arrange the slaughter, but make the customers pay for the slaughter and butchery on their own. This is probably the more "legally accurate" way to go about it.

In our own case we presold pig halves and wholes, had customers sign a contract and specify "large" or "small," and pay a 50% deposit. We would presell a full trailer-load of pigs, 15 in all, then take them all to a USDA slaughterhouse that did an excellent job and charged a very reasonable price. It was cheaper to do this than arrange for on-farm mobile custom slaughter, and it was quicker to dispatch that many pigs. We then paid the slaughterhouse a small fee to deliver the carcasses directly to a custom butcher shop that we identified as being of the highest quality and the most amenable to working with our customers. We gave our customers suggested cutting instructions but directed them to call and make arrangements directly with the butcher. Once we received the final carcass weights of each animal from the butcher shop, we assigned each half carcass to a customer by tagging its leg and would e-mail the customer with a final invoice based on the hanging weight plus slaughter costs. The customers would pay the butcher separately and be responsible for picking up their final packaged, boxed meat.

Once we got this process down, this was an easy, lucrative way to sell around 50 pigs a year, but we felt there was an upper limit to this method of selling because not enough people have the money and the freezer space to buy meat in bulk. If 50 pigs were all you wanted to raise, this would work fine. We discovered, however, that finishing around 300 pigs a year was what made sense for our land, feed storage, and profitability goals. So we had to devise other ways to process and market the rest of the pork. Having more than one outlet to sell your products is a good idea anyway—diversity brings resilience.

FURTHER PROCESSING AND ON-FARM PROCESSING

Transforming your raw ingredients into something that can last longer or be more ready-to-eat can be a powerful moneymaker for your overall farming business. Processed foods can open up new markets that were never open before, allow you to better serve existing markets with more diverse products, extend your season, and provide a host of other advantages. However, they can also add significant cost, logistics, record-keeping, and other administrative burdens to your already taxed management (probably you!). Some call these products "value added," which means more than just raising the price to consumers. It may be adding other values, such as convenience, improved nutrition, or creating more local jobs. I discuss the concept of "value added" in more detail in chapter 13, but here are a few value-added/further-processed food items to consider building into your business and to get your creative juices flowing:

1. For **dairy farmers** on-farm bottling is gaining in popularity. Farmstead cheesemaking is making a comeback, too. On-farm bottling costs might not be too high, especially in states that don't require pasteurization. Cheesemaking facilities, on the other hand, are very capital intensive and can be very hard to cash-flow because of the long aging periods of cheese, although you can make soft, fresh cheese, too; chèvre, for example. All dairy products require extreme attention to record-keeping and food safety.
2. For **animal producers** on-farm slaughter and butcher or just slaughter might allow you to have more control over your meat processing. Probably the most expensive and legally cumbersome of on-farm processing ideas, it may be possible with state or federal financial assistance. More and more regions and states

are beginning to address the lack of abattoirs for meat producers. If you want to keep things a little more simple, just developing on-farm butchering/smoking/curing without the slaughter may be an option, particularly if you have nearby slaughter facilities that meet your needs.

3. I recently met a young man who is building a USDA-approved butcher shop on his stepdad's farm to make sausages, salamis, and other charcuterie. He will buy USDA-slaughtered pigs from nearby farmers. I met another farmer who is building a mobile poultry-processing trailer that will be state-approved up to the federal poultry exemption limit—20,000 birds a year. The only reason he is making it mobile is that he is leasing his farmland and does not want to invest in a brick-and mortar-building on land he does not own.
4. For **fruit and vegetable farmers,** an on-farm commercial kitchen will allow you to make a wide variety of products, from fruit leathers to veggie chips to jams, salsas, chutneys, and more. You could even prep fruits and veggies in the kitchen and take them to farmers' markets to make fresh smoothies on site! Getting your home kitchen certified is becoming easier, as many states are passing what are known as "cottage food" laws that allow a limited amount of commercial production to be made in home kitchens (check your state Dept. of Health for the latest laws where you live).
5. For **grain farmers** on-farm milling, bread-baking, pizza-making, pie-making, and so on might help you transform your grains into foods that consumers are hungry for. From 5-pound bags of flour to wood-fired pizzas, the possibilities are limited only by your imagination!
6. For **nut farmers** on-farm roasting and making nut butters, nut flours, nut oils, and more can help you differentiate your commodity from everybody else's. A couple of my favorites are maple-roasted pecans and tamari almonds!

The following farm narrative highlights a farm that took a low-value commodity and transformed it into a high-quality, unique branded food line. The Bluebird Grain Farms story shows how a young couple went from dabbling in grain production to vertically integrated production and processing, while committing to the tenets of sustainability. Sam and Brooke Lucy perfected the harvest, processing, and storage of their grains to maintain quality and add value, creating a solid business model that others could emulate.

Bringing Ancient Flavors to a Modern Marketplace

BLUEBIRD GRAIN FARMS, WINTHROP, WASHINGTON

• Brooke and Sam Lucy •

In a world of thousand-acre farms, federal subsidies, low prices, and lengthy and complicated distribution chains, Bluebird Grain Farms stands out as one of a unique vertically integrated grain operation. They farm much smaller acreage (around 280 acres at the moment), they don't receive any federal subsidies because of the varieties they grow, they earn a fair price for their products, and they control their entire grain-processing chain from "plow to package."

Owner Sam Lucy learned about grain farming by apprenticing and working for a skilled organic grain farmer in the Methow Valley of North Central Washington. Then he struck out on his own, selling his services of farmland restoration to the many landowners in the valley who owned farmland but either didn't know how to farm or didn't have time. Like many areas of the country in which large farmland tracts have been carved up into smaller, less viable chunks of land, the Methow Valley has many 5- to 20-acre "ranchettes" that are succumbing to weed pressure and in some cases losing their water rights because they are no longer being farmed.

Sam would help landowners control their weeds by planting a crop on them, which also allowed the owners to keep their water rights and agricultural property tax status. He would also help landowners restore their lands back to native grasses and wildflowers if they chose. The crop production became more and more of a focus, and the Lucys began experimenting with heirloom and ancient grain varieties that could withstand the harsher growing conditions of the region and the heavy weed pressure that many of these fields were plagued with.

With a $1,000 used combine, Sam and wife Brooke began to harvest those grains, hand-clean them, and package them up to sell to friends and family. To sell the volume they needed to create a viable business that would support their family, they have scaled up over time and added steps to their grain processing to improve the quality and move more volume through the system. More on those steps and equipment later.

The Lucys were able to buy a modest house for a reasonable price, and once their house had increased in value (as the Methow Valley became ever more expensive), they refinanced it and used that money to capitalize their farm and buy some key pieces of farming and processing equipment. The couple has also used some conventional bank loans over their eight years in business and has received a lot of support from friends and family (less financial but important nonetheless for such things as marketing support, networking, and babysitting!). Organizations such as the Seattle Chefs Collaborative and local food networks also helped them get established.

One program that provided a key boost to their business was the USDA Value-Added Producer Grants (VAPG) program. Most farms have little to nothing in their budget for marketing expenses. The Lucys were awarded a $57,000 matching grant from VAPG to pay for essential marketing elements, such as a fantastic website and online store, marketing collateral, packaging, and product development. Brooke commented that the grant application took an enormous effort on her part, and she would have been pretty devastated if they did not get the grant after that huge input of time. However, their solid business plan, growing sales, and vertical integration all contributed to their scoring well in the VAPG application process. Now Bluebird Grain Farms is largely self-financed with business sales and a little up-front operating capital through a grain CSA.

Sam and Brooke lease land from around eight different landowners and own 40 acres of their own. Their home ranch is where they live, where all of their processing equipment, grain storage, and milling takes place, with the rest of the ranch restored to native grassland. They still work with many of the landowner contacts they made while doing the farmland management/restoration work, and some of their lease agreements don't involve any exchange of money. For some landowners the service of farming, weed control, improving soil tilth, and maintaining the owner's water rights and ag property tax rates is more than enough of an exchange. The Lucys do pay for a couple of large farmland parcels, especially their best, most fertile farmland. Luckily, all of their leased land had been fallow or managed organically, so getting all of it into organic certification was not difficult.

BROOKE LUCY

Outstanding in their grain field

In addition to being organically certified, Bluebird Grain Farms is also certified by the Salmon-Safe program. Salmon-Safe focuses on land management practices in six areas: riparian area, water use, erosion and sediment control, integrated pest management, livestock, and biodiversity conservation. Being Salmon-Safe certified demonstrates the Lucys' commitment to going above and beyond organic certification and being genuine stewards of their land.

Soil health is the foundation and focal point of the Lucys' land stewardship, as well as their business decisions. "The health of our soil translates to the health of our business and health of our community," remarks Brooke. The couple takes soil samples throughout the year and conducts leaf analysis and Brix readings so they can see how many nutrients they are taking from the soil and adjust things accordingly to yield optimum nutrition from their grains. Taking Brix readings is more common in fresh produce production, especially fruits, but to do it in grain production shows the Lucy's dedication to the nutrient density of their foods. They rotate their grains with cover crops such as vetch and use seed inoculants and foliar feeding of enzymes and microbes, as well as mineral inputs to help boost the soil health. All of the straw and plant stubble stays in the field: Unlike some farmers who bale off the straw as an additional salable product, the Lucys keep that organic material on site to replenish the soil. They use irrigation sparingly, but they do irrigate all of their fields to create a more consistent, quality grain harvest.

As mentioned above, the Lucys found that heirloom and ancient varieties of grain were hardier and performed well under organic conditions. Their "signature" grain is the ancient emmer wheat (*Triticum dicoccum*), which is a predecessor of modern-day wheat and was cultivated as far back as 8000 BC in ancient Mesopotamia. It is a hulled wheat that has higher protein, fiber, and mineral content than modern-day wheats but also has a much lower gluten content (a big asset, considering the rise of gluten intolerance in America). Bluebird Grain Farms also grows a dark northern rye and two varieties of spring wheat—hard red and hard white—both open-pollinated varieties. When Bluebird first got started, they could easily sell the rye and wheat, since consumers were familiar with them, but producing more emmer was and is their ultimate goal. They have had to do a considerable amount of consumer education to familiarize people with emmer, but its superior growing and eating qualities encourage them to keep increasing its production. The Lucys save and plant back their own seed so they can maintain the incredible quality they have developed and continue to adapt the varieties to the unique Okanogan climate.

BROOKE LUCY

Bluebird employee checking the harvest

Grain has to go through many steps to be ready for sale. When I asked Brooke what basic equipment was needed, she said, "A tractor, disc, seed drill, combine, and a screened seed cleaner." Once you start wholesaling grains, you must clean them even further, getting out any potential rocks or even bits of metal that could come off your combine and contaminate your grain. For hulled varieties like the Lucys grow, you must also invest in a good dehuller, one that won't remove the bran and germ but will get the hull off. If you want to move beyond just selling whole grains, then you need milling equipment on top of that. Brooke estimated that a farmer could cobble together an adequate grain processing setup, including both new and used equipment, for somewhere between $80,000 and $100,000. Since grain prices are low (relative to many other crops) and the investment is high, a farmer really must add value to her products and focus on quality over quantity if she wants to do it on a small scale.

The Lucys also felt very strongly that they had to control the processing and have a direct relationship with their customers—not only to capture more of the retail dollar but also to control the quality and uphold the nutritional value of their foods. For example, when a customer submits a request for emmer flour, Bluebird mills it to order. That is something that most Americans have never tasted—freshly milled flour. The difference is extraordinary.

Over their eight years in business, the couple has added new components to their grain processing line, but finding the appropriately scaled equipment has not been easy. It has taken a significant investment of research and dollars to find each piece. Oftentimes, their equipment was the "lab-scale" testing equipment that the manufacturers put together to run smaller lots. Other times they have had to refurbish an "antique" piece of farm equipment that was built when farms were much smaller. The Lucys have built their grain-processing facility step by step over time, adding equipment that would improve their production, value, and quality. Those new to grain farming might be overwhelmed by the cost and complexity of putting it all together, but the Lucys caution farmers to be patient and start with only the essential equipment, scaling up as their cash flow allows.

Brooke and Sam (along with their employees) process their grains into several products, including whole grains, cracked grains, flours, and grain mixes. They try to be responsive to their customers' requests and keep an open mind about new products. They also add value by creating quality products from split or second-rate grains, which was the impetus for their flour line. Currently, Bluebird Grain Farms uses three channels to sell its products: direct to consumer, direct to retailer, and to a few select distributors for their whole grains. The direct-to-consumer sales occur through the online store, buying clubs, a CSA, and a few farmers' markets they attend from spring through fall. They only sell their flours direct because they want to be able to mill fresh to order. Despite the higher price point (about double that of conventional grains and flour, but what it *actually* costs to produce it), customers love the flavor, digestibility, and freshness of the Bluebird Grain products.

In addition to their strong environmental commitments, the Lucys are dedicated to their social responsibility. They pay living wages to their part-time and full-time staff, provide a healthy and comfortable working environment, and pay for half of the health insurance premiums of their full-time employees. They also give back to their community through farm tours for both adults and schoolchildren and food donations to local farm-to-school efforts aimed at teaching children how to cook healthy food. They are hosting a more in-depth elementary school science project on

BROOKE LUCY

Adding value to whole grains

their farm teaching kids how to conduct field research on plant decomposition using different microbes and enzymes.

Although the Lucys still would like to increase their volume and add a bit more value to what they produce, they are at a scale that can fully support their family. It is hard for them to predict what the market will do, especially given the unique niche varieties that they grow, but customer demand continues to increase as more and more conscientious consumers find out about Bluebird. Costs of production, such as fuel and labor costs, continue to go up, so they constantly need to look at augmenting their revenues. They are building stronger relationships with retailers and are earning a solid reputation as a high-quality and high-value supplier of premium organic grains.

Brooke and Sam's advice for new farmers?

- It is important to not get overwhelmed by all the pieces. Focus on the problem at hand, and trust that it will lead you to the next phase. Take things step by step, such as purchasing a piece of equipment or getting your soils in order.
- Also, it is important to have an ultimate vision and be clear about how the needs and demands of the market fit into your vision.
- The marketplace is growing, and people are willing to pay the extra price for farm-direct products. The marketplace will drive you and support you in what you are doing as long as you have a quality product.
- Focus on what you have, and initially, follow the path of least resistance. Don't try to create a whole new system as a small producer, at least in the beginning. Use your current available assets: land, equipment, social capital, skill sets, and so on. If you don't have an asset or skill set, find someone who does and get her on your team.

TAKE-HOME MESSAGES

- In addition to a crop plan, create a harvest plan for whatever you grow or raise. This will help you prepare for harvest dates, set up processing logistics, help with market planning, and more. When you do harvest, take note of your yields, which you can use for future production planning.
- Stay on top of continuous-harvest crops because they will stop producing if you stop picking (beans, peas, and strawberries are good examples).
- Don't allow weeds, insect damage, and disease to take over toward the end of the life of the crop, thereby reducing your yields and quality.

- Properly sort and grade your products. Not doing a good job at this will undoubtedly piss off some customers and tarnish your reputation. Price and sell the different grades accordingly. However, be careful about putting out inferior product, even if you price it cheaper. Don't let inferior product dominate your stand, stall, or product line. For example, if you attend farmers' markets, it is a good idea to have some boxes of canning tomatoes or peaches available for a bulk price, but make sure you have a lot of other nice, high-grade product that dominates the table. Otherwise, customers will come to know you as the farmer who produces lower-quality product, and they will also always expect a cheaper price for your products.
- Harvesting during the heat of the day and improper or delayed cooling will reduce the longevity and quality of your products, for both crops and animal products. Additionally, not washing products that should be washed or, vice versa, washing products that should not be washed (such as berries that will start degrading rapidly once washed) will cause produce degradation.
- You can start small with grains and use animals to harvest them if you lack the specialized equipment to do so.
- Keep grains and beans properly cooled and stored. Anticipate lots of insects and rodents wanting to eat your crop, and prepare accordingly.
- You can "decommodify" (i.e., give it an identity apart from the mass-produced and co-mingled grains out there) a commodity like grains or beans, but you have to differentiate yourself and you have to add values in some way.
- Keep animals calm and in a low-stress environment prior to slaughter. Catch or transport them at night or when they are most calm.
- Let your slaughterhouse know that you expect humane animal handling, prompt slaughter, good eviscerating, and quality butchering. Shop around, and have backup options if you can.
- Be clear how you want your meat packaged, cooled, and stored. Maintain the cold chain even during transport. We purchased a refrigerated van for this; other farms will rent a reefer van on occasion to pick up their meat from the cut-and-wrap shop.
- Have good labeling for all products, including your brand, address, volume/weight, and any other information on the label required by law.
- Think seriously about packaging. How will it impact product quality? Will it become a source of solid waste or will it be

recyclable or compostable? How will the packaging impact your marketing and boost your sales?

- If you will be doing further processing, will you have control over the quality, packaging, added ingredients, and other factors? How will you store it until processing? How will it be stored after processing? How much inventory do you want and need? Will it be costly to store this inventory, or can you do more processing to order (as the Lucys do at Bluebird Grain Farm, milling their flour fresh to order)?
- Will you really make more money if you do further processing or create other value-added products? Think about *all* of the costs involved. Will it provide an insurance against risk, extend your season, keep your customers satisfied for longer in the year, give you a leg up on the competition, and so on?

CHAPTER 9

MARKETING AND RELATIONSHIP BUILDING

BUILDING ON THE DISCUSSION in chapter 2 about identifying your niche in the marketplace through solid market research and understanding the different sales outlets, agricultural philosophies, and pricing strategies, this chapter will go much more deeply into creating a marketing plan and building customer loyalty.

As with many things in life, there is no single definition of "marketing." One person may say that marketing consists of all the actions of identifying, developing, promoting, and selling products or services. A more lyrical definition is that marketing is the "art and science of creating, delighting, and keeping customers while making a profit and building enterprise value" (Max Kalehoff, 2011). Web and direct marketing guru Heidi Cohen (www.heidicohen.com) asked 72 senior marketing executives for their definition of marketing and received 72 different responses, although, thank goodness, there were 12 consistent themes. Here is my interpretation of these 12 themes with examples for **farming-related businesses**:

1. **Know your customers.** Remember the demographic research I suggested you do in chapter 2? This is where that research comes into play. You need to know who your customers are, what motivates them, and what they need so you can translate these insights into action that will build your business and keep your customers happy.

2. **Develop the right product.** This sounds too obvious, but if you try to sell customers a 5-pound roast when they don't have the time, money, or know-how to cook it, maybe you should think about grinding the roasts into easy-to-cook sausage links instead! Create products or services that solve your customers' problems and fill their unmet needs while still sticking to your values.
3. **Identify unmet needs.** Sometimes your customers or potential customers can't easily tell you what they want or need. In direct marketing you have the opportunity to develop relationships with your customers. Through these personal relationships and more traditional tools such as surveys, sales analyses, and watching smoke signals (keeping an eye out for patterns and trends), you can often sleuth out your customers' or potential customers' unspoken or unrecognized needs, desires, and values. You might find some customers are most motivated by the ability to cook and prepare a nourishing meal while others are concerned with staying within a certain food budget. Decent value for their dollar, health and wellness, good flavor and quality, safety, authenticity, and a sense of community are other motivations. Use this information to fine-tune your business strategy, product offerings, and marketing efforts.
4. **Get your message out.** You want your customers, potential customers, and, in some cases, noncustomers to know who you are (or what your brand is), what you stand for, and what you offer. Develop a clear and consistent set of messages and imagery and use them throughout your business and marketing efforts. This includes traditional media such as logos, signage, brochures, or advertising and social media, which will be important for twenty-first-century communication. I will expand on how to maximize social media's use further along in this chapter.
5. **Offer an experience, not a product.** In business, loyalty is usually critical to profitability. If you're like me, you feel a stronger bond with people after you've shared something together, whether it's beer at the pub, taking in a baseball game, or singing in the same choir. Strive to create experiences around your brand to help you bond with your customers; ideas might include offering unique recipes from a chef who shares your values, hosting on-farm events, or creating a contest to name your newest calf.
6. **Tell your story.** Develop a compelling narrative of how you started, what your values are, why you chose farming as a vocation, and what makes you special/unique/quirky. But do all

this in about one paragraph! That is about how long people's attention spans are these days.

7. **Don't just talk; listen.** Engage with your consumers beyond just the sale. Ask for feedback, and where possible, use it to improve your product or service or create new ones. This might be the time to insert a little *cause* marketing, where you engage your customers around an *issue* that you both care about. Show them that you are more than a business out to make money.
8. **Make customers fall in love with you or your brand.** What makes you *love* a company? Some words that come to my mind are honesty, authenticity, transparency, consistently high quality, dependability, excellent customer service, and guarantees; they are easy to shop, stand behind their products, customize things for you, offer good value, are more than just a business, have a compelling owner's story, are committed to their values, produce a small environmental footprint, are good to their workers, are community oriented, and so on. Why will people fall in love with you?
9. **Develop customer trust.** Trust develops from a variety of factors, including being authentic, being dependable, providing good value, standing behind your products, providing safe food, even admitting your mistakes or challenges. Whatever you do, don't fib or tell people what you think they want to hear. There's no better way than lying to break the bond of trust.
10. **Create awareness and desire.** Marketing ultimately sells stuff for you. People need to know you exist and find you relevant before they'll be willing to part with their almighty dollars. You'll need to figure out which marketing activities will best help people understand why your products are so good and stimulate their desire to make that first, second, and third purchase. This is especially important if you are considering spending money on a marketing campaign; consider how each investment will most economically sell your products or services.
11. **Everything you do matters.** To get repeat business you of course need first to have made your customer happy. You also need to be at the top of his mind or at least under consideration when he's ready to make his next purchase. The good news is that it's not as expensive these days to remind people of the wonderful things you have to offer. In today's "always-on" mentality, you are marketing even when you think you aren't: when you are on vacation, out on the town, attending church, at a PTA meeting, chatting with other farmers on Twitter about this year's hay, and

so on. Does that mean you have to wear a branded hat or uniform all the time? No, but what you say and how you act will be judged as an extension of your business. And to be honest, it may not hurt to run errands with a magnetic sign on the side of your car or a stack of business cards in your wallet!

12. **Drive profits.** If your marketing efforts simply drive sales but don't generate profits, you may need to revisit your marketing plan. For example, if all of a sudden you have hundreds of people showing up at your farm to buy piglets because they read an article in the paper, but your profits are in selling sides of pork and not baby pigs, then that marketing effort did not drive profits for you. It may not make any economic sense for you to sell piglets.

    Here's another example: You sponsor a local radio station that in turn announces your name several times each hour as a new underwriter of their station. All of a sudden your voice mail is full of charities and other radio stations asking for your donations. Instead of driving profits, you just drove a bunch of people to think your business is the next big philanthropy worth tapping. I am by no means saying this would be the case, but you need to analyze your efforts to see if they are going to improve sales of your highest-margin products into the best-priced markets. A lot of times you can't predict how this will play out, but you can certainly be attuned to the potential impacts of your marketing efforts.

BUILDING YOUR BRAND

Before you can do any of the above, you need to build a brand. How do you build a brand? Entrepreneurs often think of a brand as a logo, but it is much more than that. A brand includes your business name, your values, personalities, communications, causes you support, and more. Pick a good, catchy business name that will suit you for the long term, no matter what enterprises your farm may include or discard. For example, if you pick the name "Happy Meat Co.," then stop raising animals and turn to vegetables, people will obviously be confused. Likewise, if your business is "Conscious Herbal Creations," people may think you are selling medical marijuana, not herbal tinctures.

If you have inherited a farm name that is no longer relevant, that isn't catchy, or that you just don't like for some reason, you can always create another business name. For example, you may be the third generation to be farming at "A. Smith Ranches," but you could create a branded meat company with a completely different name, such as "Heritage Acres Ranch," that may have more popular appeal or better convey what your values are. Your business can have more than one name, such as different names for the farm and for the marketing arm of it.

Realize that it will cost some money to register several different names, and it could cause some confusion among your customers. One farm we visited in our travels had a name of the physical ranch, a branded meat company, and another food-distribution business with a different name. I never quite figured out what to call the business, and one of the owners also had a hard time explaining to people what the differences were. Perhaps dropping one of the three names might have helped that business build its brand more effectively.

Perhaps you want to name your business after an important geographic landmark nearby, after some of the values your business holds dear, or maybe even something funny or cheeky that will be memorable. One of the best farm names I have seen in my travels (sorry, it is already taken) is "Lettuce Turnip the Beet," an organic vegetable farm in California. Just make sure the name you choose is not already taken or does not too closely resemble a competitor. This may confuse people or, worse, lead to a lawsuit. A simple Google search can help you start researching a potential name, as will checking with your county clerk's Fictitious Business Name database. Finally, make sure you check with the online US Patent and Trademark Office for any registered trademark names so your chosen name doesn't copy or too closely resemble them.

Once you have chosen a business name, create a logo. There is plenty of advice on logos on the Web, but this is what I think: Make your logo memorable, convey some of your values, and, most importantly, make it something you are proud of and willing to plaster everywhere. Along with a logo, you may want a tagline to go on or near it, especially for labeling purposes. A friend of mine, produce marketing specialist Dina Izzo, created one of my favorite taglines when she worked for a produce company called ALBA Organics. Simple yet heartfelt, it was: "ALBA Organics. Where Values Are Grown In." With all the talk about value-added foods, which are essentially processed food products that command a higher price, Dina was trying to convey that values were inherent in the effort of the organic farmers who grew the food.

Branding also includes conveying your social and environmental values. You can do this through your logo, your packaging, your online and print materials, your merchandising, and the causes you support. Be clear about what those principles are, and make sure your branding is congruent with those same principles, such as environmentally friendly packaging for an environmentally friendly business. I often find it incongruous as a consumer to see organic eggs in a Styrofoam carton that is toxic to make and hard to recycle. I will reach for the product that more closely "walks the talk."

DEVELOP A MARKETING PLAN

Before you start spending any time or putting any money behind your marketing (and sales) efforts, I recommend that you create a road map, also known as an annual marketing plan. After all, you wouldn't just randomly choose a breed of cattle or sheep to raise or buy seed before you know how much you plan to grow. Things won't always go as expected, but if you have a clear plan, it will be far easier to adjust when things change. While you may assign just one person to develop a plan, I recommend that at some stage you include all of the owners, partners, business team, and possibly even your nonbusiness staff in the process. Each person will bring different ideas and perspectives that may improve the overall plan and specific details. You also want to ensure that all members of the team are on the same page and working toward the same goals. A good marketing plan format to use is as follows:

Vision/mission. Your big, hairy, audacious, long-term end result or achievement. It should be in line with the holistic goal you formed in chapter 5, but for your marketing plan it should be the end result of the marketing efforts you lay out in the marketing plan. For example, "To become the world's most famous heritage tomato producer"; "To successfully and fully transition to a grass-finished beef program vs. cow-calf production by 2022"; "To feed 100 families in my community with everything they need for a good diet and, in turn, be fully supported by them."

Goals. Your marketing goal for the year in words. For example, "Launch a meat CSA program including beef, pork, lamb, and duck." You can have multiple goals and multiple objectives under each goal; just make sure they are all congruent with each other.

Objectives. Your specific financial or other measurable objectives for the year, such as revenue, profit, volume, market share, and growth rate, related to the goal above it. For example, under the meat

CSA goal above, a specific objective might be: "Sign up 200 CSA customers and generate $250,000 in CSA revenue by January 2014."

Strategy. Your game plan to reach your marketing goals and objectives. This is the "meat" of your marketing plan. The marketing strategy section should include information about:

- Product—your product(s) and services
- Price—what you will charge customers for products and services
- Promotion—how you will promote or create awareness of your product in the marketplace
- Place (distribution)—how you will bring your product(s) together with your customers.

There are several core marketing strategies, and the ones you choose will depend on the stage of your business. These include drive trial, drive repeat, drive switching, increase "basket size," achieve cost savings. Driving trial for an *existing* product means a new customer makes his or her first purchase of that existing product. Driving trial for a *new* product or service means an existing customer *or* new customer purchases that new product or service. Driving repeat, on the other hand, means your customer makes more than one purchase or comes back for seconds or thirds.

Driving switching means someone who already belongs to a CSA has now joined yours instead. If you were selling trucks, this would mean you convinced a Ford truck driver to switch to a Dodge. Increasing "basket size" essentially means that the average purchase price from each customer goes up, say from $20 a week to $25 a week. You might achieve this by offering volume discounts or inspiring your customer to add pork to her vegetable or fruit basket. Make sure you keep profitability in mind—you probably don't want to have your customer spend more but for lower-margin items.

As a general rule if you have a new brand or product, you will likely focus on getting new customers. If you have an existing customer base, you are likely to focus on getting your current customers to buy more or encouraging people to show more loyalty to your products.

Tactics. Specific activities that help you achieve your objective(s) and that are consistent with your strategy. If the strategy for your new meat CSA is to drive awareness and trial (let people know you exist, then get them to buy from you), having a local food blogger feature your new program in a blog post might be a good tactic. Offering a

10% discount to customers who get another customer to sign up may also be a good tactic to support your strategy. In the next section I will review several other marketing tactics appropriate for farm businesses that you may want to include in your marketing plan.

COMMUNICATING WITH CUSTOMERS (NEW AND EXISTING)

There are a number of ways to get the word out about your farm and its fabulous products, including the following.

- **Social media.** These are defined as "web-based and mobile technologies used to turn communication into interactive dialog" (see Wikipedia). For me social media are a tool and a tactic that can be used to help your business, but it takes some finesse to use them well. These different media will change over time as popularity ebbs and flows for each of them. Use whatever is popular at the time. Right now this includes Facebook, Twitter, a website, and a blog. More on social media below.
- **Free website listings.** There are quite a few websites that act as directories for farmers and ranchers. Most allow you to list your farm name, a short overview of your farm and products, and contact information. They often allow a link to your website, blog, Facebook account, or other site. Some even include a feature that allows your customers to rate your farm or its products. These directories can be great for finding new customers.

  The only caveat is that you have to keep your listings updated. Write down all the sites you have listed your farm at, including the user name and password, so you can easily make updates. This is especially important if your product line or prices have changed, if you are adding or deleting an event or listing additional selling locations. Some of the best farm listing sites right now include:

  - www.localharvest.org
  - www.localdirt.com
  - www.realtimefarms.com
  - Any statewide local food guides, both web-based and print
  - Regional farm marketing sites such as www.food-hub.org for the Northwest or the various state Market Maker sites

(http://national.marketmaker.uiuc.edu/) available throughout the Midwest and the South.
- Also try to get listed on food review sites such as www.yelp.com

- **Free media attention.** You might be surprised how eager journalists, bloggers, photographers, radio and television hosts, and others in the communications world are to learn about your farm and its products. They are often looking for things to write or tell about, things that appeal to their respective audiences. Why not see whether they'll write about you or have you as a guest on their radio program? A good way to start is to identify local media, study their work (read their articles, sites, etc.), and consider what about your farm or products or story might be of specific interest to them and their readers, listeners, or viewers. You can write press releases about new things you are doing, such as launching a new CSA program or offering grassfed beef for the first time, then reach out directly by e-mail or phone to your targeted list. You can also invite these folks to visit your farm or try your products.

  One idea that I think would do wonders for your farm is to have a press-only farm tour, complete with a barbecue or a lunch featuring ingredients from your farm. Let the press see, smell, touch, and taste your farm; then release them back into their world to talk you up in their respective fields.
- **Tastings.** Offering people the opportunity to taste your products is a great way to inspire a purchase or to convince someone to feature your farm and products in a story. Look for creative ways to get people to taste your product, such as chef taste tests, hosting in-store demos, and sending food samples to well-respected food bloggers. If you are able to sell your products on a national or regional basis, by all means reach out to people who have a wider audience.
- **Respected restaurants.** Getting your products into a couple of well-respected restaurants might do wonders for your business because many people look to chefs as experts on finding unique, best-quality ingredients. Keep in mind, however, that selling to restaurants is not for everyone. Indeed, I am not convinced that selling to restaurants should be a large aspect of your business because most nonchain restaurants are fickle in nature and typically order low volumes.

  I do believe, however, that if you can get your name placed on the menu next to the ingredients you provide in a well-executed dish and get the chefs to talk you up (to their waitstaff, to their customers, to the press, etc.) it may provide an excellent dose of "buzz" for your reputation. Soul Food Farm in California is a good example of this phenomenon. After farmer Alexis got her pastured broiler chickens into the venerable Chez Panisse, other restaurants were clamoring for more birds than she could possibly produce.

- **Perks for frequent buyers.** One strategy to grow your business is to inspire your casual customers to become more frequent users and to reward your converts for their loyalty. This could include occasional product sales, coupons, bulk discounts, access to limited edition products, discount cards, free cooking classes, exclusive farm tours or dinners, cookbooks (that feature your ingredients, of course!), and more. One farmer I know will occasionally snap a photo of one of her best customers and post it on her farm's Facebook page along with a generous dose of praise, sort of a "Customer of the Month" program. In our last farming business, we used to include coupons at the very bottom of our e-newsletter as a reward to those who not only opened the newsletter but actually read it to the bottom. Some farmers send an automatic coupon to any new customer signing up for their e-newsletter or on their Facebook page. Coupons can also be used effectively to sell more of a slow-moving item or boost consumer purchases. For example, you could offer a "Buy 3 lbs. of rutabaga and get the 4th lb. free!" promotion or offer 10% off to customers who spend more than $50 at your stand. Get creative, folks!
- **Food events.** There are established food festivals and new ones popping up all over the country such as Taste of (insert your town/city name here), county fairs, and other gatherings of foodies, eaters, and influencers. Consider participating! If you can't attend yourself because of time, money, or logistical reasons, consider partnering with a chef or restaurant to at least feature your ingredients. Ever entered your produce or wine in the county fair? Ever entered a BBQ competition with your pork and beef? These kinds of events will give you an opportunity to meet hundreds, if not thousands, of potential new customers.
- **Third-party endorsements.** Display awards, certifications, and testimonials on your website, blog, print media, and wherever you sell. If you haven't won any awards, get a good customer to nominate your farm for one or two! If you win ribbons at the county fair, display them! If your butcher tells you your beef

carcass was the most beautiful she has ever seen, ask if you can quote her on your website or in your brochures. Whenever you get a compliment, ask the person if you can quote him, if he can "Yelp" you or post a positive review on your Local Harvest page, and so on. Capture those words if you can!

- **Branded merchandise.** You can also promote your brand by offering or selling T-shirts, hats, mugs, reusable water bottles (stainless steel or at least BPA-free), canvas shopping bags, coolers (to put your product into!), pens, handkerchiefs, bottle openers, stickers, and more. When you offer branded merchandise, the people carrying or wearing your brand help advertise for you. The key to choosing the right items is to pick things that people will actually use, not things that will end up in the trash or that are environmentally degrading.

  A friend of ours once made us a bunch of buttons with our old logo on it. Unfortunately, not many people wear buttons anymore, so most of them probably ended up pinned to a bag or sitting on a desk somewhere. The point is the buttons were not out there in the community being seen by lots of folks. We needed more effective merchandising. Eventually, we did get an awesome T-shirt design that people loved and wore proudly. Instead of our business paying for the advertising, our T-shirt customers would pay us above and beyond the cost to print the shirts, then go on to advertise for us whenever they wore them. Those are the kind of win-win situations I want you to try to cultivate in your marketing efforts.

WORD-OF-MOUTH MARKETING

For farms and small businesses that don't have a huge marketing budget, nor want to, word-of-mouth marketing (WOMM) is usually the most effective and cheapest way to build your business. But how do you maximize WOMM? Here are a few ideas I have learned from successful farmers and other savvy business people:

- Be **contagiously excited** about your farm, your practices, and your products. If you are depressed about your lackluster farm sales, don't let people see it! If you are worn out from working too many 12-hour days with no break, assign someone else in your family or management team to be the happy, smiling face of your business.

- **Delight your customers.** What will make them happy, get them to smile, or send them into a delicious food coma? Find out those things and focus on them.
- Think about why your customers would want to **tell their friends** about you. What makes you worth talking about? Is it some amazing flavors, colors, and smells of your food? Do you have an intriguing/inspiring/off-the-wall personal story? Are you offering some crazy good deal on something? Did you provide above-and-beyond customer service at some point?
- Go to the **influencers** who are often at the center of word-of-mouth marketing. These may be the food writers for your local paper, food bloggers, chefs who like to talk a lot, or other "well-networked" farmers and food businesses. They may also include unconventional influencers such as medical professionals (I find acupuncturists and chiropractors often are big supporters of sustainable food), sports coaches, fitness trainers, leaders of parents' groups, even local politicians or celebrities (we had a couple of city council members who bought from us regularly, and boy, did they talk us up!).
- Turn your casual **customers** into regular **users**, then hard-core **converts**. Those converts will then spread the gospel of (*insert your farm name here*). This is probably the most important tactic to increase sales locally. Those converts will become your evangelical force, your cheerleading team that will get more and more people coming to get a piece of the goodness you are selling. They will also become your greatest asset when it comes to building **social capital and good will** around your farm, a topic that I will explain later.

SELLING STRATEGIES

So now that you are an expert at marketing, how do you go about selling your food or other agricultural products? There are a variety of sales venues that I discussed in chapter 2; namely, wholesale, direct to retailer or restaurant, and direct to consumer. I am going to delve a little deeper into the sales strategies that I find most profitable and rewarding:

Direct to restaurant or retailer. Most restaurants and retailers purchase supplies and ingredients from middlemen such as brokers and wholesalers. This in part is because it's convenient. Only one truck shows up to deliver the goods, and there's just one invoice. Middlemen can often offer lower prices, too, because they purchase in bulk (so they get volume discounts from their suppliers that they can pass along to the restaurant or retailer). If you can eliminate brokers and wholesalers and deal directly with the restaurants and retailers, such as grocery stores, you stand a chance of making significantly more money. While a wholesaler might pay you 40 to 50% of the retail price, direct to retail should get you closer to 70% of that retail price.

Another advantage of selling direct to a restaurant or retailer is that you can deal in smaller volumes, unique varieties, or seasonal variations. Wholesalers aren't typically equipped or willing to deal with such complexity. If you have a crop of spring lambs in the late spring but not at other times of the year (they wouldn't be called "spring" lamb then, would they?), why not sell them when you have them to someone who will appreciate their fresh flavor, such as a quality restaurant that celebrates seasonality? Seek out the restaurants and caterers that thrive on changing their menu according to what's regionally available during peak seasons. They will be much more likely to appreciate and know what to do with your seasonally available ingredients than a restaurant that has the same menu year-round.

If you do manage to get your farm products into a food retailer, one thing to keep in mind is that the retailer is making space for you on a shelf or in a cooler. That often means nudging out some other product. Most retailers are loath to manage complexity, and switching things in and out on the shelf equals complexity. They are likely to be less enamored with you over the year if your product availability goes up and down. I don't know why retailers are used to the seasonality of some products, such as citrus or seafood, but seem to understand seasonality less for other things like eggs and meat.

We ran into this problem when we fought hard to get our eggs into a local retail chain, only to have to pull out in the winter when our production declined. Although we still had some egg production, we had to direct that production to our more lucrative farmers' markets, especially because we had Egg Share customers that had prepaid for their eggs. Our less lucrative wholesale accounts were the first to get eliminated when production slowed. The stores don't like to be the last on your priority list. Can this situation be made different? I think it can. One idea for working with retail stores interested in securing good, local products is to have the stores forward-contract with producers to get guaranteed supply. I have also heard of some stores offering production loans to a number of farmers to help them scale up and better meet the volume demands of the store.

How do you best serve restaurants and retailers? In two words: routine and reliability. Try to call, e-mail, or fax in your weekly order sheet on the same day at the same time. Find out what day and time of the week they prefer to do their ordering and take delivery. Make your order sheet easy to understand with obvious pack sizes and prices for each item. Once they put in an order, deliver it to them on time, in the correct quantity, and provide them an accurate invoice. You have to be professional to cultivate and sustain these relationships. Attempt to meet with them at least once a year (more frequently, if the relationship is new), perhaps in your slow season, to do some planning and troubleshooting and to gather feedback about the packaging, volume sold, and other pertinent matters. Spending time building trust will also build loyalty.

If you feel that the store or restaurant is not doing a good enough job promoting and selling your food, talk to them about strategies for promotion. You may need to provide more education to their staff, who are, after all, in the best position to influence shoppers. How about offering the grocers produce or the dairy team a tour of your farm? To drive sales you could also do some in-store tastings, provide recipe cards, or create some nicer-looking point-of-purchase promotional materials that they can use. If it's a restaurant, ask them nicely to put the name of your farm on the menu, or at least listed on a chalkboard or somewhere else prominent in the restaurant. Otherwise, they are not promoting your farm, which is a loss both for the restaurant and for the marketing of your farm.

Farmers' markets. There are thousands of farmers' markets across the country, and there are new ones sprouting up each year. It seems that every downtown business association or community group wants a farmers' market in their neighborhood. This can be a boon for farmers. I think selling at farmers' markets can be a valuable slice of your sales pie, particularly for beginning farms or those with smaller volumes to sell. Even commodity-crop farmers can benefit from farmers' market sales, as you will see in the narrative of a rice-farming family below. The high growth in the number of farmers' markets can also be a giant frustration, however. Some areas are getting so saturated with markets that sales are suffering, while other towns don't have the foot traffic to support viable markets that are worth the farmer's time.

Farmers' markets are great for connecting with your customers, building trust and confidence in your products, and gathering market research. If you or a family member is craving more social time, selling at a farmers' market is a great place to do that. Many farmers don't have the time to sell at farmers' markets. The good news is it does not have to be the actual farmer standing there for four long hours. It is usually quite easy to find a person craving some part-time work or who loves to eat your food and wants to be paid in product! Just make sure you provide her good training, show her how to set up your stand, and give her ongoing feedback on how she could improve sales. At TLC Ranch, we gave bonuses to our employees when they made certain sales targets, such as an extra $20 for bringing in over $1,000 at a market.

Before committing to sell at a market, visit it a couple of times to observe the foot traffic, see which vendors are doing well, and, if you are confident enough, ask a couple of patrons what products they would like to see added to the market. If it is primarily a tourist market, you may not be able to sell much of what you have, such as frozen meat or greens that must be cooked. Likewise, if it is mostly a locals market, shoppers may not be interested in expensive value-added products that perhaps tourists would be more attracted to buy as gifts or souvenirs. Nonperishable items probably do better for tourist markets, as well as foods that are packaged in containers under 3 ounces (most airlines won't let you bring any liquids or jellies in containers larger than that). When visiting the market, see whether anyone is offering what you plan to sell. If not, ask other vendors if anyone used to sell those items and what they think about your joining the market. You might learn some very valuable information on what to do and not do based on their input.

Since I have attended farmers' markets for about 10 years, I have a little more advice concerning them. Make sure you set up an inviting display and make it eye-catching, colorful if possible, and abundant looking. A vegetable farm I worked for would put small empty boxes inside their baskets, put a little piece of cloth over it, then lay the vegetables on top of that. This made the baskets look full and overflowing. Have clear pricing signage that can be viewed from at least 10 feet away, because a lot of customers wander down the center aisles first, scouting out products and prices. Display some pictures or even video (a tablet computer like an iPad works great for running video clips) of your farm so that consumers have a better understanding of and appreciation for what you do. Showing what you grow in pictures also helps build their trust that you are producing what you have on display.

Use a pop-up tent or umbrellas to keep shade on your products and try to keep them as fresh and as cool as possible during the day.

We would keep most of our eggs stored in coolers sitting under the shade and would pull out six or eight dozen to put out on the table at any given time. That kept most of the eggs quite cool. (Most states require that you keep your eggs at a temperature of 45°F even during the market. Be very sure to know your state, local, and market-specific regulations, and follow them strictly.) If you sell vegetables or culinary herbs, you can often keep their stems submerged in water to prevent wilting, or you can keep a bottle of potable water that you spray your crops with periodically during the market to keep them cool.

For meat you have to devise some system to keep it cold at markets. Some people use ice blocks in containers, then put the meat on top of it. We kept ours in coolers with frozen water bottles that we reused over and over again (saving thousands on ice costs). Our meat cuts were displayed prominently on huge signs, and when customers asked to see certain cuts, we would pull them out of the coolers. Still, I have seen farmers put their meat coolers up on their tables and allow their customers to open up the coolers and rummage around for the cuts they are trying to find. The main drawback with allowing your customers to handle the meat packages is that excessive handling can ruin the vacuum seals on the bags or cause punctures in the plastic, plus more warm air gets into the coolers and can cause the temparature to rise over the course of the market. To avoid that kind of loss, we kept our meat behind us in coolers.

Traveling in Italy, my husband and I witnessed small, mobile butcher display cases at farmers' markets, usually elevated on a mobile trailer with some sort of generator for power. If farmers' markets are going to be a main sales venue for you, it might make sense to make this kind of investment in your visual display and cold-storage capacity. I would just caution you not to make any huge investments without testing the markets for a while and building a core customer base. If you later on decide to drop farmers' markets because they just aren't selling enough for you, you don't want to be stuck with a bunch of specialized, expensive equipment that you can't transition to other uses or that are difficult to sell. This is true for any aspect of your farm—try to make *flexible* investments.

CSAs. The concept of community-supported agriculture (CSA) is that your customers become investors in your farm, sharing in both the bounty of the farm and the potential risks. In a traditional CSA arrangement, the farmer solicits customers to purchase a "share" of that year's crops, with customers often paying up front in a lump sum. The idea is that the money comes early to the farmer—in the winter or spring—so the farmer has his operating funds for the season. The farmer also knows how much volume to plant based on how many shareholders he has for a predetermined length of CSA season. Some CSAs choose not to share production risks with their shareholders but instead commit to buying product from other farmers to supplant their own, should they have a crop failure.

CSAs might vary from 12 weeks short in colder climates to 50 weeks long in others, but in all, CSAs provide a seasonal mix of farm-raised foods to the consumer. For a consumer joining a CSA is a great way to learn how to eat seasonally and stretch one's culinary imagination. CSAs usually distribute the weekly boxes through neighborhood drop-off sites, such as one customer's house or place of work. They usually allow on-farm pickup as well. A minority of CSAs actually do home delivery to all their customers, which is more common in the Buying Club model described in the next section.

One successful example of a traditional CSA is Genesee Valley Organic near Rochester, New York. In this CSA the shareholders go so far as to get together to plan the farming season, pick out crops and varieties, and determine pricing based on the farmers' goals, costs of production, and what the shareholders agree is fair to pay. This intense involvement of consumers as "partners" in the business is certainly not for everyone, but it is good to point out that Genesee Valley Organic is one of the oldest CSAs in America, and the farmer-owners not only have a good quality of life but also have built equity in the farm and their retirement savings with this unique arrangement.

Some CSAs provide more payment options to their shareholders, especially to make their programs more affordable for lower-income families. You could offer monthly payments, sliding-scale payments based on income, or even accept food stamps (Electronic Benefit Transfer or EBT) for part of the cost. Some CSA farms, such as Fair Food Farm (described in chapter 1), offer CSA work shares, in which volunteer labor is exchanged for a share of produce. Green Gate Farms in Austin, Texas, has a sponsored-share program in which they solicit their shareholders to pitch in extra funds to help sponsor shares for low-income families. They also hold a variety of fund-raisers to pay for the sponsored shares.

The beauty of CSAs is that they build community around your farm and can be used as a vehicle to do so much more, such as helping to feed the less fortunate. One CSA in California (Live Power Community Farm) actually secured their farmland at affordable terms when their CSA customers rallied to form a land trust to purchase the farmland development rights, allowing the farmers to buy the more reasonably priced land they had been farming and nurturing for over a decade.

CSAs have a reputation for being tough to manage. I should point out that much of the administrative burden that keeps many farmers from delving into CSA arrangements, or that propels them to drop a CSA as a sales mechanism, can be solved through some of the new web-based programs available today. CSAware is a useful one designed by the founder of LocalHarvest.org. With CSAware, shareholders can sign up for shares online, make payments, let you know when they'll be out of town, and order any special items you offer, such as canning-quality produce or value-added products. As the administrator you can see and manage all of this information, set what goes into the boxes each week, manage your drop-site information, e-mail some or all of your members, and print out harvest lists, box labels, and more.

Buying clubs/food subscription services. Buying clubs are similar to CSAs but differ quite a bit in spirit and in practice. They are not about sharing the risk, and they usually don't ask for the full-season price up front. Instead, customers pick what they want weekly or monthly and pay as they go, or some iteration of that. Some buying clubs deliver directly to the home (sort of a home shopping service), while others organize group drop-off locations such as churches, schools, or even businesses. Buying clubs usually offer members a higher degree of customization, like a grocery store, such as adding on breads, cheeses, meats, or eggs. One meat-buying club in California called The Foragers even offers California-grown crawdads as an option!

Buying clubs aggregate products from multiple producers, which is the model used by The Foragers and another club called Greenling in the Austin, Texas, area. Although CSAs sometimes purchase product from other farmers, buying clubs do this as the business model. They are often run by distribution businesses that do no growing themselves, as is the case with Greenling.

There are pros and cons to operating a buying club versus participating in one just as a vendor of products. Running a buying club, as The Foragers model does, means that the farmer-owner can dabble in different enterprises but still have a variety of other producers to purchase from to create a diverse selection of foods that appeal to customers. Because the farmer is taking on the marketing costs and marketing risks, he also receives a margin on the other products, which can offset the costs of marketing and transportation. Likewise, running a buying club carries less production risk because it is more of a pay-as-you-go system, rather than full payment being made up front for a full season of food.

On the flip side, as a farmer selling to a buying club, this may provide the flexibility needed for you to get your business established and perfect your enterprises without the level of management required to run your own buying club or CSA. However, you may lose much of that consumer connection, as well as a percentage of the profits, if you simply sell to a buying club. Just like any other wholesaler, a buying club will probably pay you 50 to 60% of the retail price so it can make its margins.

Traditional CSA farms increasingly have to compete with these buying club distribution schemes, which offer more flexibility and customization than a CSA. While some may see that development as good and encouraging of friendly competition, others see buying clubs as watering down the farmer-consumer relationship, skimming more money off the farmers, and challenging the notion of "local." One buying club based in Washington State, for example, is now setting up shop in Idaho, California, and even Alaska. Obviously, when the food has traveled over multiple states, from California to Alaska or Washington to Idaho, the food they are selling is no longer local.

Likewise, their consumers will have no chance to interact with the farmer growing their food because they are hundreds of miles away, which is a lost opportunity for education and awareness. However, if you are a beginning farmer or only grow a few crops or animals, you might be interested in partnering with an existing CSA farm or buying club to sell your products without having to create your own program. We did exactly this with our eggs. We partnered with a large, long-standing CSA that wanted eggs for their shareholders. It provided up-front operating capital for us and gave us an ensured outlet for a set volume of eggs.

U-pick and farmstands. If you like having people out to your farm, then organizing a U-pick operation or establishing an on-site farmstand might be perfect for your business. U-pick (also referred to as "pick your own") has risen and fallen and rerisen in popularity around the country as more families are trying to save on their food costs, enjoy the pleasures of working outside, or store food away for the winter and lean times. On the LocalHarvest website, I counted over 1,016 farms that used the term "Pick Your Own" to describe how they sell farm products, and another 2,457 that used the term

"U-Pick." This represents nearly 3,500 farms nationwide, and that's on just a single farm-listing website. Another site that appears to be growing in popularity is called PickYourOwn.org, which lists U-pick and pick-your-own farms by state and county along with what is available to pick and times of operation.

My own family used to pick apples to make homemade applesauce—a favorite food of my brother and me as kids. Twenty years later, now that I have my own homemade applesauce-loving child, I have rediscovered the value of picking apples. I think a lot of other families are coming back to U-pick, too, and I think there is tremendous opportunity to bring more U-pick farms across the country. For farms that are having a hard time finding picking labor, switching over to a U-pick operation is a logical solution. U-pick farms also dovetail quite nicely with efforts to promote rural tourism and agritourism. Some of the key considerations to evaluate if you have the right location, crop mix, staffing, and so on include:

- Road access and parking
- Signage
- Bathrooms
- Staffing needed
- Teaching proper harvest techniques to U-pick customers
- Safety and liability
- Potential permits needed from your county, township, and state and other regulatory requirements
- Planting the varieties and successions so that you will have enough product to meet demand

You can always start really small and casual, such as having a U-pick day or two for your best customers, family, and friends, and expand over time as you gain experience and work out any kinks in the system.

Farmstands can also be a great way to introduce people to your farm, allowing them to connect to the sights, smells, and fresh flavors you have to offer. As with U-pick operations, if you are well off the beaten path, you have to consider whether or not people are going to make the effort to get out to your farm. If you do have a farmstand, make sure you create an eye-catching display, you have good signage, and that the products are kept fresh. You can also put out bins of lower-grade or wilted product that you offer at a discount or free for the taking, but don't allow lower-quality products to dominate your offerings because it could negatively impact your reputation.

To create the diversity of foods you want to fill out your stand, you could purchase a limited amount of product from other nearby farmers or have them bring it to you and sell it on consignment at your stand. Having good quality and diversity of product will keep customers coming back for more. Some key considerations when thinking about creating an on-farm farmstand include:

- Road access and parking
- Signage
- Bathrooms
- Staffing needed
- Safety and liability
- Potential permits needed from your county, township, and state and other regulatory requirements
- Product quantity, diversity, and keeping it all fresh

You could choose to have both a farmstand and a U-pick operation. For example, you might decide to let your customers harvest the crops with the highest labor costs. Peas and strawberries come to mind as high-labor crops. You could also plant a "cutting garden" in which customers can go out and make their own flower bouquets or cut their own culinary herbs. The rest of the crops could be already harvested and sitting at the farmstand ready for sale. Misty Brook Farm in Massachusetts (profiled in chapter 5) did something like this. Their farmstand held all the dairy products, meat, and most of their vegetable crops. The strawberries and cut flowers were offered to their customers as U-pick.

Some people believe you have to be close to major population centers or be diversified fruit or vegetable farmers to really benefit from direct marketing, but that is simply not the case. Farming businesses around the country are proving that notion wrong. Bluebird Grain Farms in Washington is a good four hours from the greater Seattle region. The Foragers buying club is organized by a farm that is five hours from San Francisco. Butterworks Farm resides in the second least-populated state in the union (Vermont) with the tiniest capital city (Montpelier). Massa Organics in California is not only three hours from most of its customers, but they raise mostly rice, not exactly the first crop that comes to mind for direct marketing. I think the Massa story below shows that virtually any farm in any place can develop a brand, find appropriate markets that value quality, and build relationships with consumers. Their marketing savvy has been key to the family's success.

From Monoculture to Many

MASSA ORGANICS, HAMILTON CITY, CALIFORNIA

• Greg Massa and Raquel Krach •

Harvesting the 95th crop of rice must make a family proud. That is precisely what the Massa family did last year on their family-owned farmland in the Sacramento River Valley of Northern California. However, just like many multigenerational farms, the youngest generation was not content to do things the same way. Fourth-generation farmer Greg Massa, along with his wife Raquel Krach, have transitioned a third of the family acreage to certified organic production, about 220 acres in total. Not only is the couple demonstrating that growing rice organically is possible but that it's also possible to direct-market this commodity, build strong consumer support around it, and integrate other crops and livestock into the mix.

The main enterprise of Massa Organics is organic brown rice. In rotation with the rice, they grow organic wheat, organic hay, and, occasionally, organic alfalfa. Greg is looking for an additional cash crop to put into the rotation because he prefers his fields to have a two- to three-year break between rice crops. Rice culture can create quite a bit of soil compaction, also, weed growth needs to be broken up—thus the need for a longer rotation. He is considering experimenting with oilseeds, such as safflower, but will have to identify an oilseed processor that is also organic within a reasonable distance of the farm.

On some of their better-draining soil, the couple planted almond trees about five years back. The trees are now producing good nut crops, although Greg is quick to admit his rice-farming background did not exactly prepare him to be an orchardist. He is still making improvements in how he nurtures the trees. Just recently, in the winter of 2011, Greg and Raquel added hair sheep to their orchard management system to help with weed control and to provide fertility. They are still in the experimental phase, observing the feeding patterns of the sheep, keeping an eye on potential tree and irrigation system damage, and figuring out how best to fence them in and protect them from predators. Once they get their flock size up, they can consider marketing the lambs as feeders or as retail cuts of meat, turning what was once only a cost of production (weed control) into a new income stream.

Likewise, in 2011 the family added pigs to the mix, raising them on fallow rice fields or on the field edges to help control some of the perennial weeds that plague organic rice production. Turns out that pigs love cattail tubers and Johnson grass rhizomes, two key weeds in North American organic rice production. Since broken rice is an inevitable by-product of rice production, the Massas can now turn it into more profitable pork instead

RICK KRACH

Harvesting the mature rice

of selling it as animal feed. In 2012 the family will begin to direct-market their organic pork, which will make their current farmers' market stands even more profitable. Raquel is especially fond of adding animals to their list of enterprises and is taking a lead role in their management. This is in addition to helping with the behind-the-scenes business management and taking care of their amazing brood of five children.

When the Massas grew their first organic rice crop, they knew they wouldn't be able to wholesale it to stores. That market was already locked up by some much larger organic operations with marketing budgets bigger than most farms gross in a year. They thought they could market rice by partnering with existing vegetable CSAs, providing organic rice as an add-on option. That didn't take off as much as they had hoped, mainly because CSA farmers are already notoriously busy managing complicated operations. So they tried selling at a farmers' market in a nearby town. Although the volume sold was not massive, getting the full retail price for their rice was appealing and held promise for this avenue of marketing.

Turns out they were the first rice farmers in the country to sell rice at farmers' markets. (The entire rice commodity system is set up to aggregate rice from hundreds of farmers and sell it all over the world, while obscuring the identity of the grower.) The Massas, on the other hand, were going to grow, mill, and sell their own Identity Preserved rice

RAQUEL KRACH

Pigs attacking a ricefield berm.

RAQUEL KRACH

Weeds are annihilated by the pigs.

direct to the consumer. Doesn't seem like an earth-shattering event, but it apparently was in the world of commodities. They "decommodified" a commodity. Bloggers, food writers, and chefs took notice once the Massas started selling at the San Francisco Ferry Plaza Farmers' Market, one of the busiest markets in the nation. Now, more than 10 years later, the family is selling its rice at 15 farmers' markets a week around Northern California, as well as wholesaling a portion to restaurants and schools. Through their relationship with school food service provider Revolution Foods, the Massas' organic brown rice is nourishing 30,000 to 40,000 schoolkids each week.

At the Massa Organics market stand, you can find their employees selling 2-pound bags of brown rice, bulk wheatberries, whole-wheat flour, roasted or raw almonds, and almond butter. In the future they hope to add retail cuts of pork and lamb to their booth if the markets allow them. One solution to making farmers' markets a financially viable sales venue is to continually figure out how to add value to the markets a farmer attends. The Massas are doing this by adding niche products that few others have. Each summer they also raise a small batch of ducks in their rice fields, selling whole ducks to those eager customers who reserve them early.

Ducks and rice production are a common combination throughout Asia, but California rice culture makes it a little tricky. Instead of transplanting 6-inch-tall rice plants into the paddies, as they do in Asia, California rice farmers actually seed their rice by plane. You have to wait until the rice emerges above the floodwaters before you can put ducklings in the field. By that point the weeds can be quite thick, and the small ducks aren't as effective at controlling them. Then the ducks have to mature before the rice sets seed because they will eat the seed. So getting the timing right makes for a challenge, but one the Massas are willing to spend a few years perfecting. They know that the addition of ducks can provide weed and pest control, natural fertility, and an added income stream. It turns out that Greg and Raquel's backgrounds in ecology are coming in handy as organic farmers of today—this ecological awareness can be seen throughout their operation.

Greg and Raquel spent their twenties studying biology and doing research in tropical ecology in Costa Rica. Raquel was looking at trophic levels specifically, which is about the food webs that exist in nature. But they left academic research, craving a more hands-on approach to ecology that could be found in agriculture. The couple moved back to Greg's family farm—a place he never expected to return to. Now they take that understanding of food webs to their farm ecosystems. For example, a rice field is essentially a pond ecosystem. While a conventional rice farmer might be focused on advancing a single species (rice), the Massas are trying to encourage other species to thrive in that pond environment. For example, crawdads can live and thrive in a rice paddy without doing damage to the rice crop. A farmer could actually harvest and sell some of those crawdads if she were so inspired.

Ducks and other wild waterfowl can provide natural weed and pest control,

RAQUEL KRACH

Bringing the ancient tradition of raising ducks and rice together in California

while depositing their rich manure into the water. Modern rice varieties need quite a bit of nitrogen to grow, and this nutrient is often a limiting factor in organic rice production. Therefore, any nutrients you can get from waterfowl are welcome.

As ecologists the Massas also view their farm from a holistic perspective. Cosmetic stuff such as a few weeds here and there or piles of branches are not something they worry about—these "untidy" patches provide food and cover for beneficial insects and other helpful organisms that contribute to ecological balance.

Growing organic rice can be tricky, especially in the first few years after transition from conventional production. The soil microbiology is very low as a result of years of chemical fertilizer use and compacted, anaerobic soils. Getting enough organic nitrogen during the crop year is hard; Greg uses a combination of winter cover cropping and animal manures to get enough into the rice plants. Because weeds can't be controlled with herbicides in organic production, and because rice lives in a flooded environment in which you can't use soil tillage, weed

RAQUEL KRACH

Livestock bring some fun to a traditional grain farm.

management can be tough. The Massas use a multipronged approach to weeds, including crop rotation, thick cover crops, preseeding tillage, animal impact, and even a little bit of nature. Their winter cover crop attracts lots of migrating birds and rodents, both of which help to eat some of the weed seeds.

In 10 years Greg and Raquel have gone from a 20-acre experimental plot to 220 acres now in organic production. Much of their profits have been reinvested back into this expansion, but they have finally reached the scale at which they won't need to be adding acreage. The farm they live on is completely organic, while their parents still farm rice conventionally some 20 miles away on a separate ranch. Profitability should be better now that expansion is no longer needed. The only thing Greg and Raquel will be adding now is increased diversity of crops, some new livestock in rotation, and increasing the value earned for their existing crops.

Not all value-added products pencil out—the couple discovered this when they tried to turn their whole-wheat flour into tortillas. Customers loved the tortillas, but they were highly perishable and cost quite a bit to have processed. With the cost of processing and the loss involved with selling a perishable product, they lost money on the product. Luckily, their giant pile of frozen tortillas (they would freeze whatever returned from market) went to the happy bunch of pigs they're raising. Now whenever the Massas have excess wheatberries, they can feed them to their pigs, which is another way to add value to the grain without nearly as much risk. Turns out meat protein is a lot more valuable than stale tortillas!

Marketing is an area that Greg and Raquel have excelled at. They started with a good logo—one that is colorful and memorable and that conveys their values. It includes a picture of their innovative rice-and-straw-bale house, a rice field, and some waterfowl. They also have a well-designed website and a blog (which Greg says is too much work to keep up), as well as a popular presence on Facebook and Twitter. Massa Organics takes online orders through their website and also through their Facebook page. They try to bring as much of the farm as they can to their customers, most of whom live several hours away, through frequent photos, video clips, and stories of what is happening on the farm. For those that can visit the farm, the Massas host a variety of farm tours for schoolchildren, seniors, and other groups.

Although Greg and Raquel don't attend the farmers' markets themselves anymore, they try to keep their market employees up to date with what is going on at the farm. Since there is a limit to how much rice people eat and it can last for a long time, the Massas have strategically added other niche products to their market stand, especially ones that people go through much quicker, like a bag of raw almonds or a jar of creamy almond butter. They are creating a whole new group of brown rice addicts; many of their customers have told them they didn't know brown rice could taste so good and that they no longer eat white rice.

Greg and Raquel's advice for new farmers?

- Don't try to do everything at once. Especially if you are converting cropland from conventional to organic, don't try to do it all at once. Don't stretch yourself too thin, either. Build it up over time. Have patience.
- Don't be afraid to try something new or different. Be creative, as it doesn't hurt to try new ventures on a small scale.

DEALING WITH CHANGE AND COMPETITION

No amount of business advice and planning will work if you are not prepared for change. You have to keep your farming model flexible enough to respond to changes in the marketplace, new competition, and even altered weather patterns. This is not easy as a farmer. You may have invested a ton of capital into specialized equipment and infrastructure, specific kinds of land, and employee training. You may have developed efficient marketing methods that get your products into the hands of consumers in a way that makes sense for your bottom line. You may have done specialized breeding of your own locally adapted crops or livestock breeds. But all of that could change if your main buyer finds a new, lower-cost source, or feed prices skyrocket out of your control, or consumers decide to buy from a corporate buying club instead of an individual CSA program like yours.

Doing a SWOC analysis (as described in chapter 5) every year during your marketing plan review will help you foresee some of the changes that could come your way and plan for change. You need to constantly be thinking about retaining your existing customer base and attracting new customers. You also should consider ways to turn potential competitors into collaborators, if at all possible. For example, if a new buying club is starting in your community and you are afraid it might draw away business from your farmstand or your CSA, could you at least become a main supplier of produce to that buying club so they are providing another sales outlet for your products? Could you convince another quality meat producer to refer his customers to you when he sells out of meat, and you will do the same for him? We each do better when we all do better.

STRENGTHENING CUSTOMER RELATIONSHIPS

Apart from being a savvy marketer, you have to think about ways to strengthen your relationships with your customers. This will also prepare you to endure new or increasing competition or other market changes. You may want to start by thinking of consumers in a different light. The word "consumer" itself to me connotes a superficial relationship built on simple transactions and the idea that a person buys something, gobbles it up, then spits out some waste at the end. I'm pretty sure that is not the type of relationship you want to build, because they are fleeting, often degrading, and not sustainable for the long term.

To be a farm with a future, you should think of your customers as partners and part of your network of social capital. Social capital, just like financial capital or human capital, is something you can nurture and draw on when you need it. Social capital is the value of social relations and the role of cooperation and confidence to achieve collective or economic results. Unlike other forms of capital that deplete over time from use, social capital actually increases with use! If you ignore your social networks and think you can do it all alone, you will reduce your social capital and the chance to turn your customers into loyal *partners*. There are many manifestations of social capital; I will give you a few real-world examples that relate to running a farming business.

- A new farmer in California suffers a horrible wildfire that burns up half of her pastures and several of her chicken coops and kills thousands of young birds. Because of her dedication to nurturing an engaged customer base, several of her most loyal customers organize benefit dinners at area restaurants to raise funds for getting this farmer back on her feet. A bunch of these customers also come to her farm and rebuild a couple of her chicken coops in a day. She recovers and continues farming with a renewed sense of optimism.
- A new farming couple in Texas must move their animals quickly because of a fast-moving wildfire spreading dangerously close to their ranch. Friends, neighbors, and customers with land offer to trailer their animals to safer locations to ride out the fire. People act quickly and come to get the animals out. The fire passes through the property and burns much of the pasture, but no animals or humans get hurt. When the fire has passed, all the animals are returned safely to the farm.
- A young farming family in Massachusetts needs to find a place to live near their farm's fields, but the options are limited. They find a house they can afford to buy, but they can't qualify for a conventional mortgage because of their self-employed status. Ten willing and trusting customers offer to finance their home purchase through individual private term loans. Now the family has a home base, a place to store all their equipment, and a better chance of staying in an area with rapidly rising real estate prices. The investors have a reliable place to put their money, which earns them better interest rates than most mutual funds. All are satisfied.

Research has shown that farms with an identity and more direct relationships with their customers have better survivability. Even if you choose to wholesale some of what you produce, try to maintain some direct market relationships, or at least preserve your identity through your wholesale supply chain. The more you invest in building your social relations, the more social capital you will have to carry you forward through thick and thin.

SOCIAL MEDIA AND ADVERTISING

There are a growing number of farmers who use social media on a regular basis. Greg Massa, featured in this chapter, is a great example. He engages his customers almost daily with Twitter and Facebook updates, photos, and short videos, sometimes even while riding on the back of his tractor! Despite being a good three hours away from his main markets, he can transport his customers to his farm virtually, keeping them engaged and purchasing his farm products on a regular basis directly through his website and Facebook webstore.

Another farmer, Nathan Winters, profiled in chapter 1, developed a strong web presence before he even started farming. Now in his first year of commercial production, he has a great online following that he can engage to become CSA subscribers or even private investors. Social media helps you build the social capital that I described above, although it should be combined with more tactile relationship-building experiences, such as food tastings, farm tours, and one-on-one conversations.

Social media is often thought of as a free way to advertise. First of all, time is money, and you will have to dedicate some time on a regular basis to keep your content fresh. If you don't have the time or enthusiasm to do it yourself, this may be a great way to involve a volunteer, CSA shareholder, or teenage child who is tied at the hip to his computer. Also, to really maximize your social media content, you may want to hire professionals for services, such as a blog or web designer, someone to design a great e-newsletter template, or even a professional photographer to take some stunning photos for your website or Facebook page. You can also buy ad space on Facebook, something Greg Massa of Massa Organics did. He was pleased with the results and attracted new customers to order via his Facebook webstore.

A blog might be easier to manage than a website because you can readily update the content without having to pay a professional web designer to update the html code every time you want to make a change, such as including a new event or changing a price. There are templates that facilitate easier administration of a website without your needing to pay a professional, but you usually have to pay a monthly fee for that ability. I found it much easier to trade food for a professional web designer than to bother to update our website myself. I just didn't want to spend any more hours staring at my computer screen than I already was.

Websites can vary from a content-based website that talks about your business to an e-commerce site that focuses on selling products and services people can order directly off the site. Many business websites include both content and product sales; the Massas do this on their website, www.massaorganics.com. Likewise, Facebook has a new webstore option available for business pages, allowing a business to sell directly through their Facebook page and regularly update products and pricing. A free Facebook application that allows you to accept PayPal or credit card payments is called Payvment Social Commerce. You must set up a merchant account with PayPal beforehand, but that is probably something you are going to want to have for a variety of reasons.

Don't start a blog in lieu of a website if you never plan to update it. A consumer might think you are out of business if your last blog update is from two years ago. Also, I have seen blogs that are so poorly organized that you can't readily find *what* they sell, *where* they sell, and other basics about buying from that particular farmer. Instead of maintaining a full-fledged blog, you could try microblogging. Twitter is a popular microblogging website where you can post anything you want in 140 characters, but you have to keep your message clear and concise.

What should you talk about in a sentence or two? Share updates about what is happening on the farm, events, new products, sales, links to articles of interest (should at least be somewhat related to food and farming), or links to recipes, or you can even post photo links to your Twitter page. Twitter takes a while to get the hang of, and it can often seem pointless because the information comes and goes so quickly. However, some marketing experts swear by it, so it might be worth a try. Just don't overdo it—I think one or two good posts a day are about all you should bother with, and post them at key times when there are lots of viewers online (probably between 9 a.m. and 6 p.m., no later). Another advantage of Twitter is that it can help you stay on top of trends and see what other people are buzzing about.

TAKE-HOME MESSAGES

- Put some serious thought into the name of your business. Will it stick in people's heads? Does it convey your values, your location, your family name? Will you have both a farm name and a brand name? Will this cause confusion? Likewise, will the name encompass all of the enterprises that you may add some day in the future?

- Try to create a quality logo, and build positive brand awareness from the start. You want people to identify you, to relate to you, to become loyal to you over the long term. Building a brand and keeping it consistent and relevant will allow you to do that. Have an artist draw a logo for you, and start putting it everywhere. Watch out, though—some research has shown that consumers can be put off when a farm looks "too polished." When this happens consumers sometimes think that a farm has "gone corporate" or is spending more time on image than on substance. You may even want to get feedback from a random assortment of your customers on your potential logo or web design to see which one they prefer.
- You are a walking billboard for your brand. Remember this when you go out on the town, attend meetings and conferences, do your grocery shopping, and so on. Build goodwill in the community for yourself and your farm, even if you don't think folks will immediately become your customers. They may over time, and they may become important in other ways. This is the beginning of your social bank account, otherwise known as social capital.
- People who like people should be selling your products. If that is not you, that is okay. Find somebody who enjoys people and is excited about your products. Show your presence every once in a while because people like to see and talk to the "real farmer," too. For example, you can attend the farmers' market once a month to say "Howdy!" to folks, talk recipes and cooking, and catch up with fellow farmers, while the rest of the time your employees, family members, or partners attend the markets. Likewise, if you distribute product to stores, restaurants, CSA drop sites, and the like, hire a friendly driver who is excited and knowledgeable about your products, growing practices, and so on for the inevitable questions she will receive.
- Don't wholesale all your products right from the get-go because the low prices received will probably not allow you to grow your business. Line up some direct markets. A mix of the two (direct and indirect) is also a good idea to diversify your sales venues and protect yourself from risk.
- Anticipate change, food fads, and increasing competition. There are more and more small farmers entering the field. People will want to emulate your perceived success. How will you deal with food fads changing? Restaurants opening and closing? Lower-cost producers or those with less integrity attempting to take market share from you? How will you keep your customers

coming back for more? Can you reward loyalty in any way? Periodically, ask your customers *why* they buy from you, and make sure you continue to deliver those qualities they are looking for.

- That said, don't run yourself ragged trying to follow every food trend, new crop, fancy breed, and so on. Don't think you have to grow everything to satisfy your customers. Can you partner with others to offer more diversity instead?
- Don't diversify into too many unprofitable ventures, such as running around dropping off $20 orders at a dozen different restaurants. Also, don't think you have to have "loss leaders"—you are not a discount grocer.
- Competition can become *cooperation.* Think seriously about how you might be able to do this. For example, you could combine forces with another vegetable farm that has a slightly different climate from yours to create a more robust CSA with both cold- and hot-season crops. A neighboring pig farm could supply you with the piglets that you want to grow out every year, and maybe you could help them market their pork, too.
- Don't assume your customers are yours for life. Be flexible; add products, services, experiences,and so forth that make sense for you and help to keep customers coming back for more.
- If you start to think of your customers as *partners*, the whole experience might change, and you will build loyalty. Invite them to be more than a monetary transaction. Let them be part of your story and you part of theirs. Enrich each other's lives. Build community around your farm, and you will not only create customer loyalty, you will also be creating a sort of insurance policy against risk and building up your social capital. Who knows, maybe that one customer who likes balancing her checkbook using QuickBooks could become a volunteer bookkeeper for you. Or that auto mechanic could help fix the transmission on your tractor in a pinch. Or that web designer could help spiff up your website, drawing in more sales traffic. You never know who is out there itching to get more involved!
- You don't have to let people onto your farm all the time or even make sales direct from the farm, but allowing the occasional farm tour or customer appreciation BBQ day will help seal the bond between your customers and your farm. Sadly, farmers who wholesale all of their products or don't have strong, personal connections with their customers don't get as much support during disasters, economic downturns, and other difficult times.

The research shows this. A wholesale dairy farm in Vermont lost a barn to a fire. Nobody took notice. A CSA farm in Vermont that lost a barn to a fire had over $300,000 in donations, local chefs held numerous benefit dinners, and a local brewery even made a special beer to raise funds—more than they needed to rebuild their barn. Why did this disparity happen between these two farms? Relationships and social capital.

- This is all about food, right? Help your customers become better cooks. Make sure you cook and eat your own food, too. This is extremely important. I can't tell you how many vegetable farmers I have met who call themselves "meat and potato" men—they don't eat the very vegetables that they produce! How can they explain how to cook things? How can they convey the taste, aroma, and texture of the wonderful things they are growing? You have to believe in what you are producing and convey that excitement. Eating it is the best way.
- Don't ignore social media, and likewise, don't expect it to create miracles for you, either. You need to be more than one-dimensional to your customers and potential customers. People need experiences—tactile, sensory, and auditory—to relate to your farm and your products. Use social media to remind them, but use in-person experiences as much as possible. If your customers all live four hours away, pick one weekend during the best weather to invite them for a campout, barbecue, wildflower walk, gleaning day, and so forth where they can experience your farm firsthand. Then keep reminding them of who you are via e-newsletters, websites, Facebook updates, new YouTube videos, and other media.
- Don't just toot your own horn while on social media. Use social media to toot *other* people's horns, provide education or references, or even engage in political discourse, if that is a part of your business model and consumer base. Consumers that see you as more than a business may become more loyal. Seventh Generation products found this out when they began engaging their customers in campaigns against toxins, and their sales increased as a result. It can be tricky from a marketing standpoint, but if you have a social mission, engaging your consumers in social and political acts will remind them of your brand and mission in different contexts and may also demonstrate your integrity to them. Keep it honest, though. Don't, for example, advertise that you are donating to wildlife reserves while simultaneously using pesticides that are highly toxic to aquatic life in your local waterways. Consumers will want to see you "walking the talk."

CHAPTER 10

RECORD-KEEPING AND REGULATORY COMPLIANCE

UNFORTUNATELY, AS A FARMER you probably won't have the income to hire an office assistant anytime soon. Maybe you have a bored teenager you could put to work or a retired parent living nearby that could help with the office responsibilities. Stacks of paper don't sort themselves magically, applications don't fill themselves in, and production records stored in your head don't just leap onto the screen of your computer. Record-keeping and regulatory compliance are serious work and becoming ever more complex for farmers. The world expects us farmers to feed everybody safely, steward our natural resources, provide good work environments, and take on all the liability and risk of doing so because they don't really want to pay us a living wage for our efforts. It's a tall order, one that many veteran farmers or would-be farmers have given up on. What are some time-saving and sensible record-keeping strategies to help you keep track of it all while reducing your regulatory risk? There is no "best" record-keeping system for farming businesses, but whatever you devise it should:

- Be user-friendly (for all who use the information, including your employees)
- Be flexible enough to provide information in a variety of ways (can you send it to a banker, to a government agency, to a buyer, to your employees?)
- Provide accurate and necessary information

You don't have to use a computer-based record-keeping system, especially if it's going to create a lot of pain and suffering for those inputting the data. If you are more comfortable carrying around a small notebook with you everywhere you go, that's fine. Some employees, particularly those who are less technologically savvy or those working exclusively in the field, could use small notebooks, and your office could have computer-based record-keeping programs to transfer the field information to a central location. Either way, it is not so important *where* you keep track of your records (although don't let your records get rained on or eaten by the dogs!). However, it is crucial that you and your team make a commitment to keeping your records **up to date** and **accurate** over time. Make sure you properly train everyone who is supposed to be keeping track of records. If some of your employees have limited English skills, translate the record-keeping template/data into the language they feel most comfortable in—that will improve the odds that they'll be able to complete the records accurately.

Now, let's go through the different parts of your business that require organization and tracking of records: production, licenses and certifications, food safety, labor, and safety. Financial record-keeping will be discussed in chapter 11.

Production records. You should develop some way to keep track of what you are planting or raising all the way through how it yields at harvest. We discussed crop plans and harvest plans in chapter 8, but this goes beyond just that. For crops this would include such things as soil tests, rotation history, preplant inputs, seed sources, seeding rates, irrigation applications, fertilizers, labor inputs, pest populations, harvest dates, yields, and, ultimately, where that product is sold.

For livestock it would include where the animal came from, the date it was born or arrived on your farm, breeding lineage, feeding rates, field rotations, health records, production records (such as how many eggs it produced, milk yield, litter size, and so on), processing records (date and location slaughtered, days hung, carcass quality, meat yield and grade, and the like), and sales records.

For value-added food products, you usually have to keep records on ingredients, recipes, temperatures, sample testing results, and sanitation, as well as where you sold the product. All these details can add up and eat up a significant amount of your time in tracking them. Devise a few different strategies to keep track of this information efficiently and properly archive it.

There exist several computer programs (both software and web-based) that help farmers keep track of various data, such as AgSquared for diversified crop farmers or Ranch Manager for livestock producers. For complicated CSA administration there are also web-based programs that can track your customers, payments, delivery routes, harvest quantities, and so forth. One such program is CSAware, described in more detail in chapter 9. Many offer free trials so you can see if you like the program. You can also develop electronic spreadsheets with weeks or months across the top and data details along the side instead of using web-based programs. Below is a sample production record-keeping format with the first five weeks of the year of a small pig enterprise:

PIGS DETAILS	WEEK 1 (1/1/12-1/7/12)	WEEK 2 (1/8/12-1/14/12)	WEEK 3 (1/15/12-1/21/12)	WEEK 4 (1/22/12-1/28/12)	WEEK 5 (1/29/12-2/5/12)
Piglets purchased, age, from? (male/female)					100 piglets, 7 weeks old–Fair Haven Farm (45M/55F)
Piglets farrowed and survival (male/female)			45 piglets born	40 still alive	40 still alive (22M/18F)
Sow name, farrow date, and (#) of piglets surviving			Sow: T14, 1/15/12 (10) Sow: T11, 1/16/12 (10) Sow: 03, 1/16/12 (11) Sow: 08, 1/18/12 (14)	Sow: T14 (10) Sow: T11 (9) Sow: 03 (10) Sow: 08 (11)	Sow: T14 (10) Sow: T11 (9) Sow: 03 (10) Sow: 08 (11)
Farrowing issues			Sow: T14, no problems Sow: T11, one piglet crushed Sow: 03, one piglet runt died Sow: 08, two piglets crushed, one piglet runt died		
Castration dates					21 males, 1/30/12; kept 1 intact male from T14 litter

You can also write a lot of this information down on a paper calendar. Keep one in your vehicle to record farm production details and one hung in your packing/processing area to keep track of yields. For example, we kept a calendar in our egg-washing room where we would write the total number of eggs received, the number broken and rejected, and the final number packed in dozens. With this information we could look at production trends and also understand what our loss rate was.

It is not enough to keep track of your inputs just based on the expense receipts. You should also be keeping track of how much each input application weighs, or its volume and the date applied. For crops this might include how many pounds of gypsum, how many yards of compost, how many inches of irrigation water, and so on. For animals, you could keep track of your daily feedings by number of hay bales, pounds of hay, or pounds of feed. To do this accurately, you should also know how many total animals you have so you can come up with a feed per unit calculation.

If you raise laying hens, for instance, you should attempt to count the hens once a month; then you can roughly figure out how many pounds of feed are going to each bird. Overfeeding or underfeeding your hens can have dramatic consequences on their rate of lay, egg size, and overall health. Although we did not do this regularly (only once a year), in retrospect we should have counted our birds monthly. We had issues with jumbo eggs and hens' uteruses prolapsing in their second year of lay, which partly can be attributed to overfeeding. We thought the issue was because of their breed, which is certainly one factor but not the only issue. I am certainly glad I know this information now, if we ever decide to raise commercial laying hens again.

Keeping track of this level of detail is normal in large, highly efficient farming operations; they make their money off the pennies that add up by way of volume. However, many small- to medium-scale farming operations, especially the highly diversified ones, tend to worry less about the minute details of production inputs and yields—but that complacency can lead to drastic problems. For example, if you continue to breed with an animal that has had poor litter numbers in the past or produced offspring that had poor weight gain and lower-quality carcasses, you are investing a bunch of time and money into a poor breeding stock.

Instead of getting sentimental about a breeding female and hoping she improves, turn her into bacon or ground beef at once! If you are going to invest in breeding stock, you must get the best genetics and mothering ability you can afford. Pay attention to their performance and cull them after the first poor reproductive event (unless the problem was truly the fault of management, not the animal). Keeping good production records will help you notice these things sooner rather than later. Similar advice holds for crops. Keeping track of the details, such as seed quality, germination rates, and yields, will help you fine-tune your operation over time and improve your profitability.

Licenses and certifications. You should make photocopies of any applications you send in for licenses, registrations, and certifications (or you can scan them into your computer if you prefer to reduce paper consumption). Keep them filed according to the name of the agency, such as your state's department of agriculture, the USDA, or the certifying body. Order the applications in each file, with the most recent in front and each previous year behind it. Once a file gets too thick, switch to an accordion file or a drawer of a filing cabinet. If a certification also requires that you keep track of production inputs and sales outlets, such as organic certification requires, make copies of those things and put them in the certification file, with the original receipts going in your financial record-keeping files, too.

For example, if you purchased an organic fertilizer, you should keep the original receipt in your "Expenses" accordion file under the Fertilizer/Amendments tab and another copy in your organic certification file. That way, when you have your annual organic inspection, you will have all of your required documentation at your fingertips. I also keep copies of any communications I may have had with the agency by printing up the e-mail conversations, copies of mailed letters, or a written description of the phone conversations I may have had, with the date, whom I spoke with, and his complete response. This can be essential if you have to document a conversation you had with an agency staffer.

For instance, when I tried for months to get the FDA to approve our new preprinted egg cartons and finally got the right person on the phone who then approved it, I wrote down that person's name, the date, and what she said. If anybody had asked us when or how we got egg carton approval, I had documentation of that verbal authorization. Just as with your financial records, you should keep all of these records for the previous five years. This is a requirement in case of an inspection or audit.

Food safety. Although most farmers and ranchers I know are intuitively safe producers of food, you should plan to keep some records regarding food safety. I am not going to discuss the various government-sanctioned or private third-party food safety standards because they vary by state, by crop, and by whom you sell to. It is up to you to decide what system of standards you will use, if any (if you sell direct or grow lower-risk crops, there are few if any standards, other than common sense). At the minimum you should document the different parts of your production chain and what food safety data is important to collect for each step along the way. For many this reads as preplanting, planting, harvest, post-harvest, and sales. For animals, it would be the same steps: before the animals arrive, while you are raising them, processing, storage, and sales. Develop a list of protocols for each step and how you will keep track that they are followed.

For instance, perhaps you want to have each field's soil tested every year, as well as the irrigation water. If your protocol is to test annually, then you will need to record when those tests were taken, by which lab, and what the results were. If you had to make a decision based on those results, what was the decision? Say another food safety protocol you have is that your employees have to wash their hands after using the restroom. Although you can't sit in the bathroom and watch them wash their hands, you can document that you trained them and that you keep the water tank, soap dispenser, and paper towel dispenser full at all times.

Labor records. For all employees, contractors, apprentices, volunteers, and even family members, you should keep private files on each of them. Make sure you keep copies of their I-9s, 1099s, job offer letters, agreements, and so on. Also keep copies of any safety trainings they have attended, as well as any disciplinary communications you have made with them. Additionally, create some way for employees to keep track of their hours by crop/enterprise and activity. This labor-hour tracking will be indispensable for you to do some of the financial planning described in the next chapter. Also keep copies of their pay stubs, performance reviews, raises, and any other documents related to their compensation in the same file.

Human safety. Over the past couple of years, I have known one farm worker who was killed by electrocution while moving aluminum irrigation pipes, a farm owner who lost an arm while using an augured oil press, a young farmer who died in a tractor rollover accident, and another young farmer who died when a truck backed over her on a steep hill. These stories really shook me and made me particularly concerned about farm safety. I am especially anxious for younger, first-generation farmers who may not have grown up on farms and learned their way around farm equipment. The last two stories above were precisely that scenario—young beginning farmers with only a year or two of equipment experience.

What does this have to do with record-keeping? Everything. A good safety record-keeping system will not only document your safety efforts, it can also point out any gaps you may have around safety training. Make sure you document each safety training you provide your farm team, and have attendees sign their names, indicating that they attended. For employees with less experience, make sure you give them ample training and have them shadow more experienced equipment handlers before they start operating the equipment themselves. Also document any personal protective gear you give out and safety enhancements you make to your equipment, such as tractor roll bars or chain-saw kickback guards. If you witness employees not using their personal protective gear, document that, too. Make sure you follow up the documentation with a conversation with the employee, and document it, too. If they get hurt not using the protective gear you told them to utilize, then at least you have the records to show that you did your due diligence.

LIABILITY PROTECTIONS

Be sure your farming business is protected from legal liability. This includes your physical farm, your labor, and your products or services. Workers' compensation insurance should protect your employees, and a comprehensive general liability insurance should cover your physical farm and the products or services you provide from it. It is important to talk with your insurance agent about all the details of your liability policy. Does it cover on-farm accidents, animals escaping and causing problems elsewhere, your doing custom work for a neighbor, and any problems with your products, such as microbial contamination or tampering?

Product liability insurance protects the business from claims related to the manufacture or sale of products, food, medicines, or other goods to the public. It covers the manufacturer or seller's liability for losses or injuries to a buyer, user, or bystander caused by a defect or malfunction of the product and, in some instances, a defective design or a failure to warn. When it is part of a commercial general liability policy, the coverage is sometimes called "products-completed operations" insurance. If you sell food (particularly those higher-risk foods such as meat, seafood, eggs, and fresh produce), you should make sure your general policy includes this coverage.

REGULATORY COMPLIANCE

There will be federal, state, and county laws that affect how you farm and run your business. Whether or not you follow all the rules is up to you, but you should at least know the laws that affect you. Many of the laws are about proper stewardship of natural resources, protecting labor from exploitation and safety hazards, and keeping food safe at different stages of the production chain. To be a sustainable farmer, you should at least be following the spirit of these laws. A good place to start to understand federal laws is the Environmental Protection Agency's "Agricultural Regulatory Matrix," found at this link: http://www.epa.gov/agriculture/llaw.html. You can find out what your state laws are by contacting your state's department of agriculture, department of health, and department of natural resources or conservation as well. They may have a whole set of other laws that affect you.

For example, while the USDA provides an inspection exemption for on-farm slaughter of poultry—up to 20,000 animals per year as long as you sell it directly to the consumer—each state can decide to interpret that law however it wants. Unfortunately, it means that some states have adopted the flexible USDA standard of 20,000 animals, while other states won't let you kill a single poultry species on your farm without a USDA inspection. If you live along a state border, this variability may put you in a bit of a bind because you won't be able to sell state-inspected meat into another state that requires USDA inspected meat. Likewise, even counties have their own sets of laws and get to interpret the federal laws as they choose. This is especially true with county health departments—they usually have a lot of power. Try to get on their good side (or I suppose you could try to hide from them). However, it's hard to run a real, viable farming business without the regulators taking notice. Do your homework on the laws that affect you, keep copies of those laws on hand, and try to stay in compliance with them. Document everything!

Below is a story of Montana's first and only goat dairy and cheesemaking facility, called Amaltheia Organic Dairy. Because of the nature of their business, they have a whole slew of regulations they must comply with, and good record-keeping is a key component of their business. But they manage to stay in compliance and keep a pleasant sense of humor about the whole thing, while simultaneously creating a premium, organic product and maintaining a viable family farm that includes their children as well.

Of Curds and Whey

AMALTHEIA ORGANIC DAIRY, BELGRADE, MONTANA

• Melvyn and Sue Brown •

After years of working in the seed stock dairy and beef cattle industry doing embryo transfers (which is a complicated breeding practice to more quickly develop the genetic potential of superior paternal and maternal breeding lines), British native Melvyn Brown decided he wanted to get into livestock production himself, but this time with a smaller animal. A literal swift kick in the knee by a dairy cow forced him to take a different direction with his career. While working in Central America, he found that goats were a great, versatile animal for smaller-scale farms and decided to take that concept to America. Having been around cow dairies his whole life, Melvyn knew what kind of infrastructure was needed for milking animals. Goats needed mostly the same thing, just on a smaller scale.

Following stints in Central America, Texas, and Colorado, Melvyn and his wife Sue moved to Michigan, where Sue had grown up. There they had a 55-acre farm where they raised all of the food for their young children. They also raised and sold organic beef. After eight years of making it work in Michigan, the couple missed the mountains and the beauty of the West, so they started out for Montana and their goat aspirations.

With the sale of their Michigan farm, the Browns had the money to buy land in Montana. Sue taught at the local high school and earned the off-farm income for the family, while Melvyn focused on building the farm infrastructure in the first few years. The couple started out with dairy goats but did not have the money or experience yet to bottle and market their own milk or dairy products. So they wholesaled their goat's milk to cheesemakers in the Bozeman area who made a private-label goat cheese for a San Francisco company.

When that cheesemaker went out of business, the Browns had to quickly figure out what they were going to do with their milk. It made the most sense to turn the milk into value-added products that earned more income and would ship more easily. They learned the cheesemaking business from vintage cheesemakers in San Francisco and took over the old cheese facility in Bozeman. After they'd been in that facility for a few months, the state decided the facility was substandard and needed to be replaced. So the Browns decided to build a new cheese plant.

Fortunately, a locally owned bank in the nearby town of Belgrade was excited about the Browns' business plan and willing to make a loan to them. They also got funding through the Montana Department of Agriculture's Growth Through Agriculture (GTA) program, a special pool of grants and loans to fund new products and value-added processing in the state. The Browns initially attempted to build the cheese plant on their farm, but the issues of permitting, environmental compliance, and sewage became too daunting to move that project forward. A newly constructed business park in the town of Belgrade was available for a long-term lease. It was already connected to the city water system and sewage and gas lines, so it was much easier to get permitted rather than building from scratch on farmland. Another benefit of locating the cheese plant within the Belgrade city limits was qualifying for a city revolving fund because of creating new jobs. Also, the Belgrade airport sits conveniently across the street, so the Browns could distribute their cheese by air easily.

Building a goat dairy and a grade A cheese plant is not easy—in fact, Melvyn estimates the total cost is probably around $300,000 to build it from scratch. Finding the right equipment for smaller-scale operations and smaller animals such as goats and sheep can also be challenging, though Europe and Israel produce equipment designed for small dairies. Added to the capital costs and hassle of finding equipment is the unending number of regulations that you must understand and follow. The Browns had to become very familiar with the federal pasteurized milk ordinance. They are the first and only goat dairy and goat cheese plant in Montana, which means that the state regulators didn't have either the experience or the understanding of how goat dairies work. However, Montana is very supportive of its agriculture: In fact, one of its current US senators is an organic grain farmer himself (Jon Tester). The state is willing to work with the Browns because they want to see agriculture succeed.

Melvyn and Sue have their dairy and cheese plant inspected monthly by state officials, FDA officials, and the county health department and annually by their organic certification inspector. Every batch of milk that heads to the cheese plant has a sample withdrawn from it, and these samples are sent to the state lab for bacterial testing. Melvyn thinks the numbers of regulations are increasing each year: "The rule books just keep getting thicker and thicker," he says. One example of a nonsensical rule that the Browns must follow is the requirement to purchase a $7,000 antibiotic tester and regularly test all their milk for antibiotic residues. But as a certified organic goat farm, they are prohibited from using antibiotics in their milking herd, making the tester unnecessary.

The Brown family milks around 250 does, which are a mix of Saanen, La Mancha, and Alpine. With a long history in the world of livestock breeding, Melvyn pays keen attention to his breeding program, finding that F1 crosses produce the best milk in terms of butterfat and protein (F1 crosses are the first generation born of crossed purebred parents). Goats are easily adapted to the high-elevation, dry mountain climate of the Gallatin Valley. As long as they are kept out of the wind and rain, they do quite well. The goats are milked twice a day and sent out to pasture between milkings. Using portable, four-strand electric fencing, Melvyn is able to rotate his goats around their farm acreage and keep the pasture in good, productive condition. Actually, Melvyn's son does much of the electric fencing nowadays,

SUE BROWN

Melvyn and Sue cuddle a couple of baby goats.

while Melvyn handles the milking and Sue runs the cheese plant. During the winter the goats get homegrown organic hay and organic grains that the Browns raise on their land and on neighboring organically certified land.

With large predators more plentiful in Montana than in many other states, the Browns use a proactive approach that has seen only a single animal lost to predation in 12 years. The protective measures they use include llamas and alpacas in the same pasture as the goat herd, as well as one Great Pyrenees dog that roams around the perimeter. Electrified fencing is another line of defense.

In addition to goats the Browns now raise pastured pigs that are fattened on the whey that is a by-product of the cheesemaking process. They also produce and market organic compost with the goat and pig manure that doesn't get deposited directly on the fields. The Browns are committed land stewards; they have restored riparian forest along the creek that runs through the property and fenced the livestock out of that area. Wildlife such as Swainson's hawks, bats, and elk can regularly be seen sharing the land with their domesticated animals.

The name of the Brown family's farm is Amaltheia Organic Dairy. The name Amaltheia comes from Greek mythology, referring to the goat that nursed Zeus. In the story Zeus was so happy with Amaltheia that he took her horn and made the cornucopia, and he put her image in the sky as Capricorn. The name Amaltheia signifies abundance.

At the cheese plant they make soft cheeses such as chèvre, ricotta, feta, and an aged Gouda-style cheese. They would love to consider other goat dairy products such as yogurt or other aged cheeses, but a perplexing state rule won't allow them to process them in the same facility as their soft cheeses. The Browns work with national distributors such as FSA, Sysco, and Organic Logistics to get their cheeses into natural food stores around the country. They also ship cheese through FedEx or UPS and through Delta Airlines.

Amaltheia pork gets marketed within state since it's processed under Montana state inspection, not USDA inspection (which allows you to sell across state lines). At this point they don't have enough pork to service their in-state demand, so it's an area they would like to expand and perhaps do more value-added processing such as dry curing.

The Browns find markets and build customer loyalty through farm tours, social media, doing in-store demos, and attending food expos such as the Natural Products Expo in Anaheim, California, each year (luckily, the Montana Department of Agriculture co-ops the price of the Montana-Grown booth so the Browns don't have to pay it all themselves).

At Amaltheia Organic Dairy the family is constantly exploring new ways to be more sustainable. Their next step is to do renewable energy such as solar or wind and work toward carbon neutrality. Their organic farming practices, rotational grazing, and homegrown compost applications have vastly increased the carbon storage in their soils. With increased organic matter they have also seen the water-holding capacity of the soil increase, plant productivity rise, and more diverse wildlife taking up residence on their farm. They are also planting pollinator plants to increase the diversity of native pollinators that visit their farm. The Browns have recently erected two high tunnels to grow vegetables for the local community. Likewise, the Browns have their own organic vegetable garden and are quite self-sufficient homesteaders.

SUE BROWN

Amaltheia Dairy's spreadable goat cheeses

While most dairies in this country probably abhor visitors, the Browns have opened up their farm gates to guests. They provide tours to international visitors that Montana State University brings to the area, as well as to alternative high schools, senior centers, elementary schoolkids, preschools, and cheese buyers from around the country. The Browns don't charge for their educational tours, figuring that an understanding of and appreciation for what they do will lead to more sustainably minded consumers and future Amaltheia customers.

Melvyn and Sue's advice for new farmers?

- Never listen to the naysayers: If you've got an idea, go with it. It might take a little while, but the future is going to be great for organic, sustainable food. If we had listened to the naysayers about goat cheese in Montana, we would have never started.
- Work closely with everyone that supports what you are doing. Opening your farm up to tours builds customer loyalty (especially when baby goats are around—customers just eat that up!).
- Come up with a good plan for what you are doing.
- Do it in a sustainable way.
- Be 100% behind what you are doing.
- Health care starts with what we eat. The future is bright for people growing healthy food.

TAKE-HOME MESSAGES

- "Record-keeping? What record-keeping?" "I think I have a desk somewhere underneath that pile." "The dashboard of my truck holds all the records I need!" Don't fall into this trap.
- Take record-keeping seriously; make time for it regularly, and make it a priority (don't let it slip to the bottom of your giant to-do list).
- You are a farmer *and* a secretary. If that makes your head hurt or turns you off farming, find somebody who can reliably and accurately do your record-keeping for you.
- Keeping everything in your head is not record-keeping. It doesn't matter if you have memorized all the important details, such as your planting calendar, harvest calendar, sales orders, what you spent this year, or which animals got bred when. A banker, regulator, buyer, employee, business partner, and all the rest cannot see what is in your head.

- Just because you are hidden in a hollow or valley or on a middle-of-nowhere farm does not mean that regulators won't notice what you are doing.
- You can find out about regulations without giving away your name, location, or other personal information through some good old-fashioned research and a well-placed anonymous phone call. Knowledge is power.
- There are layers upon layers of regulations that may affect you, from your city to your county to the state and federal levels. Some may not apply to you, but others may. Do your homework, and keep copies of the rules that may apply to you.
- When getting regulatory information, permission, approvals, and other communications, make sure you write down people's names and the dates that you talked to them; then write down the pertinent information they gave you.
- If regulators require you to test certain products or test for certain pathogens, always take a second sample at the same exact time under the same parameters. Send that test to a well-respected, certified lab. This data may come in handy at some point.
- Understand which government agencies regulate and which cannot regulate. Some agencies, such as the National Resources Conservation Service, are nonregulatory: They are not there to enforce regulations or turn you in for a violation. This may help you in deciding which government officials to let onto your property and which not to.
- If you have integrity and are a good steward of your land, animals, and employees, you should not be fearful of government officials. If you have something to hide, you may not be creating a sustainable farm.
- Have a general liability insurance policy, as much as you can afford. Make sure it covers your products, too.

CHAPTER 11

ACCOUNTING AND FINANCIAL MANAGEMENT

I KNOW YOU PROBABLY didn't choose a farming vocation because you love spreadsheets and accounting, but it is one of the necessary "evils" of running a business. As I have said before, I want you to stay in business for the long haul—and that requires that you have a complete grasp of your business's financial performance. Once you get proficient at the number crunching, it can actually become quite fun playing with the numbers. It will also give you better peace of mind to know how your business is doing throughout the year, rather than waiting until tax time to find out.

Pulling together all your sales records and receipts once a year in March to prepare your taxes is not sufficient to build a solid business. If you wait 12 months to find out you are in the hole for $25,000, how will you dig yourself out and have the operating capital you need for your next year? Will you catch things, like the restaurant that is four months late paying your invoice or the feed bill that went up 10% in price even though you order the same amount each month? There are so many details about your farm's financial health that you will miss over the course of the year as you dump your receipts and sales slips into that shoe box. It's time to move from the shoe box (or truck dashboard) to a spreadsheet and filing system, either handwritten or electronic.

FROM SHOE BOX TO SPREADSHEET: BASIC BOOKKEEPING

When we made the change from once-a-year bookkeeping to a regular, ongoing system, we hired a contract bookkeeper. We just didn't have the time ourselves to do a thorough job, so we found an experienced small-business bookkeeper who was already a steady farm customer. She was happy to be compensated in meat and eggs rather than currency. We found that twice a month appointments with her—during which she came to our house, entered everything into QuickBooks, filed things for us, and generated reports—was a good system for us.

My husband and I would then review the financial reports at a monthly family business meeting. We paid particular attention to sales trends at our different farmers' markets and how we could reduce key expenses such as feed. We would usually pick one expense each month to research, looking into alternate suppliers, prices, free or used sources, and other research to bring our costs down.

For example, one year our truck repair expenses were unusually high as we tried to keep our main farm truck operable after 250,000 hard miles on it. It became apparent that the cost of these major repairs no longer made sense, given the value of the truck. So we made an informed decision to buy a new used truck in great condition, and we sold our old one to someone who wanted the engine. Other little expenses such as labels and printer ink added up, and noticing those things quicker (because of monthly profit and loss reports) allowed us to seek out preprinted labels in bulk for a much cheaper price per unit. We eventually discovered we could get preprinted egg cartons without labels and not only save money but also time putting the labels on each carton. That may seem trivial, but not when you are producing over 350 dozen eggs a day. Five to eight seconds a carton to affix a label turns into 30-45 minutes of labor every single day!

There are several ways you can better keep your finger on the financial pulse of your farm. **First** is having a good record-keeping system. **Second** is entering that information regularly into bookkeeping forms, electronic or paper based. **Third** is moving beyond basic bookkeeping to accounting, which is reporting and interpreting the bookkeeping numbers. **Fourth** is making business decisions based on those financial reports, then monitoring and adjusting to keep your farm profitable over time.

An excellent new resource on all these topics is the book *Fearless Farm Finances: Farm Financial Management Demystified* published by the Midwest Organic and Sustainable Education Service (MOSES, 2012). It is 250 pages of very approachable, clear descriptions of what systems you need to develop to manage your finances well, with narratives of several farmers who have taken steps toward better record-keeping and accounting. If you read this book along with Richard Wiswall's *The Organic Farmer's Business Handbook* (Chelsea Green, 2009), you will be in good financial shape. Richard helps you "plan for profits" and pick the enterprise mix that is most profitable for your farm.

If you have the time or inclination to do all your own bookkeeping, that's fabulous. However, I think bookkeeping and accounting are one of the few areas in your business where it is easy to find a qualified person to help you. Numbers people are everywhere, and they can often be quite affordable, too. They may even want to work for food. Just make sure you call a couple of their references to make sure they do good work. My only regret with our business is that we never moved beyond the basic bookkeeping; I wish we had periodically hired an accountant to help us with the financial interpretation. We relied too heavily on one report, the income statement (otherwise known as the profit and loss statement), without looking at all the other reports that would have helped us understand our financial health.

We talked a little about record-keeping in the previous chapter. There are production records, labor records, certifications and licenses, regulatory compliance records, and financial records. You should develop systems to organize all of them and, ideally, have backup copies of the most important documents. This can be accomplished when you have paper records, such as sales slips, and electronic copies once you have entered their information into a bookkeeping program. Be aware that you should always keep the previous five years of financial information in case of a government audit or investigation. Create a mouseproof archival system, such as sealed rubber boxes, to store that information off your desk and out of the mouse's mouth. We found that our old financial records became very cozy mouse nests, a problem I don't think the IRS would have appreciated should they have chosen to audit us!

Here is a simple way you might want to organize your financial records:

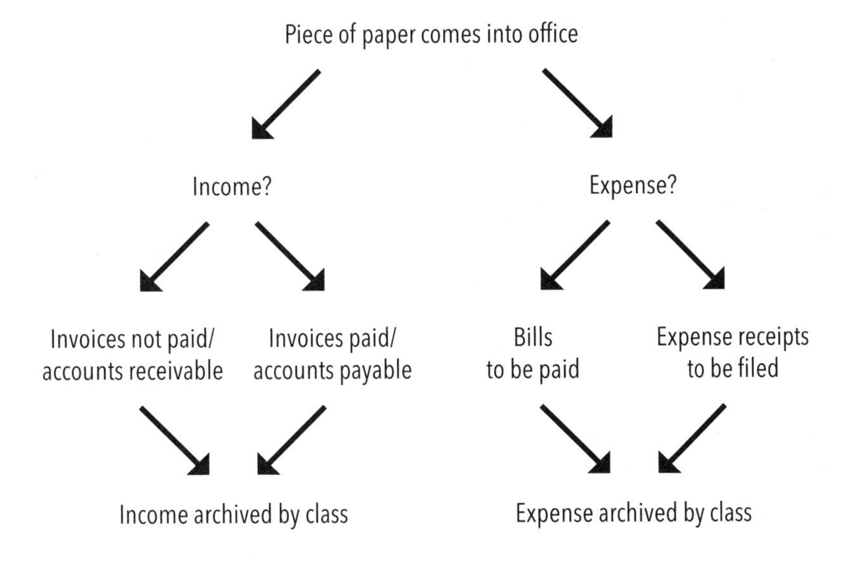

How does this flow chart work? Starting with whatever pieces of paper come into your office, you divide them into either **Income** or **Expense** (make sure you print up any invoices or receipts that arrive electronically, too). For **Income**, you have to determine whether or not it has already been paid. If it has been paid, you put it in an "Accounts Payable" file. Cash sales slips also go into that file because they are paid (such as farmers' market cash sales). If an invoice has not been paid yet (such as one from a wholesale account), drop it into the "Accounts Receivable" file. Those terms did not work for me, so I called them "Paid" and "Unpaid." Once you have inputted those invoices and receipts into your bookkeeping program, you can archive them. I used an accordion file with different labels for each class. My Income classes were: Farmers' Markets, Restaurants, Stores, CSA, Tours & Workshops, and Private/Direct to Customer. The papers would go into each section according to date, with the most recent in front.

On the **Expense** side you can divide them into bills that have yet to be paid and receipts for expenses already paid for. You can organize the bills according to the due date, with those most urgent on top. I like to write the due date in bold with a Sharpie pen on the outside of the envelope as a reminder. For expenses already paid, drop them in another box or file for your bookkeeper to enter. Once she has entered that info, you can then archive it in another accordion file labeled "Expenses" on the outside. Some businesses choose to archive expenses according to month, but I prefer to do it by *class* because it makes your life easier during tax time, when you need to know how much each expense line item was for the whole year. My Expense classes included Licenses/Registrations/Certifications; Stock (for us it was chicks and weaners); Feed; Fuel and Vehicle Maintenance; Marketing Costs (phone, Internet, website maintenance, labels, banners, and so forth); Farmers' Market Fees; Office Supplies; Land & Water, Small Tools & Equipment; Slaughter/Butcher/Cold Storage; Labor; Insurances; Seeds & Haying; Building Supplies; Miscellaneous; and Assets (intermediate and long-term).

Now that you have organized your financial files in an understandable way, you have to enter them into a spreadsheet. We chose to use QuickBooks because that was the software that our bookkeeper was most familiar with. Other programs include Quicken, Moneydance, and Peachtree. You can use Excel as well, but the benefits of accounting software might lead you in that direction instead of to an Excel spreadsheet. Some of the advantages of accounting software include being able to write and print checks, enter bank deposit details, record transactions and reconcile various bank accounts, maintain vendor and customer lists, enter bills and due dates, create invoices and receive payments, manage accounts receivable, manage payroll, and even upload your information to tax accounting programs such as TurboTax. In the beginning of our business when everything was less complicated, we relied on Excel spreadsheets and accordion files to manage our finances. However, our switch to QuickBooks bookkeeping resulted in much better record-keeping on our part, made tax time that much easier, and helped us improve our overall profitability because we had a clearer picture of our income and expenses.

ACCOUNTING LINGO AND FINANCIAL FITNESS

I am not going to go into much detail about all the financial statements and ratios you may choose to utilize to understand your financial health. There are so many different ones, and each statement could be its own stand-alone chapter; that is what the book *Fearless Farm Finances* is for. I will just go over a couple of basic concepts here.

The balance sheet. This is a financial statement that provides a snapshot of your farm's financial position at a given point in time. It works best if you adjust your balance sheet at the same time each year; for example, doing it the first week in January to account for any changes over the previous calendar year. A balance sheet includes all the things you own, called "assets," and all the dollars your farm owes to others, called "liabilities." The difference between assets and liabilities is considered to be the "owner's equity," also known as "net worth." Comparing the balance sheet from year to year will let you know if the value of your investment in the farm is growing or shrinking. This statement will also give you a total debt-to-asset ratio, which is something bankers look at when considering giving you a loan. Around a 30% debt-to-asset ratio is considered very good, between 30 and 38% is good, and over 38% is not so great.

The income statement. Otherwise known as the "profit and loss statement," this report allows you to assess your profitability for a given time frame. An income statement includes all of your revenues (incomes) at the top minus all of your expenses below that; your net farm income will be the result. "Net farm income" represents the amount of money your farm gave you over the time frame in return for the labor, expertise, and capital you invested in it. I think it is wise to look at this report monthly, quarterly, and yearly. You don't include asset expenses in this statement, or income generated from the sale of assets—those things go into your balance sheet, described in the paragraph above.

Even if your income statement shows profitability, you may not have good cash flow, and your business could cease to exist because of that. Likewise, an unprofitable farm operation can trudge along for years if it has positive cash flows. This can happen when the farmland is owned free and clear, such as with a multigenerational farm, so there are few overhead expenses. So let's talk a bit about cash flow.

Cash flow budgeting. This is a report that will help you predict the inflows and outflows of cash in the future, which helps you develop strategies for times of potential cash shortages. You should first start by looking at historic cash flow, using the previous year or two as a guide. If your business has grown, or added or dropped certain enterprises, your numbers will obviously not be exactly the same. But a historical perspective will help you develop your future cash flow projections. I like to lay it out month by month, with totals for the full year. You include farming income and expenses as well as nonfarm income and expenses because those things affect cash flow, unless your farming business is a separate tax entity (such is the case if you have incorporated your farming business and are no longer functioning as a sole proprietorship or partnership). You can use your cash flow budget to reduce your financial risk and do better planning. But remember, a budget is a projection based on your best guesses—unless you have a bunch of production contracts or outstanding accounts receivable, you won't know exactly what your income or your costs will turn out to be. Recent versions of Quickbooks will allow you to input your budget projections, which will then facilitate running budget versus actual reports on a monthly basis. This will help you see how much you vary from your budget and pinpoint specific problem areas.

If there are times of the year when cash flow is great, can some of that money be set aside or invested better to cover the times of shortage? Would a line of credit or short-term operating loan from your bank help you cover those shortages, to be paid back later during profitable times? Can you stockpile some inputs or prepay utilities when you have good cash flow to cover the times when you don't? Here's an example of this: When our egg sales were rocking in the summertime, we would purchase a pallet of preprinted egg cartons ($8,000 dollars' worth) that would last us a full year. That way in the winter, when our production slumped, we didn't have to worry about coming up with the money for egg cartons.

Although your farm might be profitable overall, you may suffer from poor cash flows, especially as a new farmer starting out. This often happens when you decide to make a bunch of capital purchases in the first few years of farming as your sales are just starting to build. It is better, if you can, to structure payments for those assets over a longer term, such as the life of that asset. If you have to pay it off too quickly, that will negatively impact your cash flow and put you at risk of failure.

This exact scenario played out on our farm, and I am thankful that I can now see what the problem was. We took out $30,000 in private loans and another $25,000 in prepaid egg shares, all of which had to be paid back in 24 months. That meant that we had $55,000 of debt (both cash and promised inventory) to pay off in only two years, and the monthly payments were too hard on our cash flow. Paying our regular operating expenses was really difficult as a result. We took on this debt load to double the size of our laying hen flock and buy three farm vehicles we deemed essential. In retrospect we would have been better off scaling up a little more slowly and financing one or two vehicles through a longer-term bank loan (for instance, 60 months) with a reasonable interest rate.

Cash flow is one of the biggest reasons beginning farmers should consider renting land for a good while before they ever think of purchasing it. If your monthly mortgage costs are too high, they will hurt your business cash flow and may cause bankruptcy in short order. In our travels we have met farmers who could not afford to buy enough feed for their animals, fertilizer for their crops, or even cover crop seed to cover the bare soil, simply because their cash flow was so poor because of too much debt repayment. That is a scenario spiraling out of control and a sustainability nightmare.

MONITORING AND CONTROL

So you've created your reports—now what? As I mentioned above, you should look for deviations, such as expense line items going up, certain sales categories going down, and times of the year with poor cash flow. Then you may need to decide to make changes in your business model. How do you make the best possible decision? Using the concepts of holistic management, as described in chapter 5, you want to get in the habit of testing those decisions against your holistic goal. Here are seven tests that I would suggest: use all seven if you want to make decisions that are not only financially sound but also in line with your holistic goal.

Cause and effect. Will this decision address the root cause of the problem or just cover up a symptom? For example, will going out and buying 200 tons of hay address the reason your cows are hungry, or is it your pasture management and recovery periods that are making the cows hungry? While you may need to address their hunger in the short term, you really should get at the root causes, or you will just keep throwing money at the problem without solving it.

Weak link tests (social, biological, and financial). Will this decision solve one or several of your weak production links? A social weak link could be a shortage of skilled labor. A biological weak link could be your soil's poor water-holding capacity. A financial weak link could be your cash flow in the winter. You should prioritize expenditures that either are *wealth generating* or *address a weak link* (or several).

Marginal reaction. Will this decision give you the most bang for your buck? What is the return on investment (ROI) of this decision? For example, you could spend $1,500 on a soluble fertilizer whose nutrients will be gone after one year, or $1,500 on a large compost application that will slowly release nutrients and feed your soil microorganisms over several years. Which provides the best bang for your buck? You may have to use a few more of these tests to make that decision, such as the weak link test above.

Gross profit analysis. Which enterprises contribute the most toward covering the overhead of the business? This test only applies if you have two or more enterprises. Say you farrow-to-finish pigs and you raise stocker cattle. The pigs require more infrastructure, breeding stock, and so on to produce, but you make 40% net profit on each one. The stocker cattle you buy at a year old and finish for around 10 months on your pastures. You don't need any infrastructure for them, and your fences were already there, but you are able to make only 15% net profit per head. Say the pigs net you $120,000 overall and the stocker cattle $30,000 overall. The pigs are contributing the most profits toward covering your overhead expenses, but they also require more overhead and long-term asset expenses. Would it then make sense to put more energy into maximizing the profits of the pig operation or the cattle? Again, don't just use one test—try to use all seven to make your decision holistically.

Energy and money source/use. Does the energy or money come from the most appropriate source in terms of your holistic goal? Is it internally or externally generated? Is it addictive?

Sustainability. Does this decision move you toward or away from your holistic goal? Of course, sometimes short-term and emergency decisions have to be made to stay in business, keep your crops or animals alive, and so forth. But try to work toward more proactive, strategic decision making that will serve you for the long term.

Society and culture. How do you *feel* about this decision? Does it lead toward the quality of life goals you described earlier? Will it adversely affect others (not only people but animals, too)? When all is said and done, if your gut reaction toward a decision is negative, then it may not be a good idea.

PAYING YOURSELF

If you are the farm owner and not incorporated, you should probably pay yourself in the form of an "owner's draw," which will be considered an equity payment. This payment will show up on your balance sheet but not on the income statement, but make sure you include it in your cash flow budget so you have the money for it. The amount you draw will be based on your family's cash needs and the profitability of the farm. Obviously, you can't draw out $2,500 a month if your farm is only grossing $3,000 a month. To keep things clean and in line with accounting best practices, you should have separate bank accounts for the farm and your personal income and expenses. You could write an "owner's draw" check from the farm bank account at the beginning of each month to deposit into your personal checking account; that way there is a clear record of the transaction and a clear separation of the business from the personal.

Use that personal account to pay for your house rent/mortgage, groceries, home utilities, clothing, entertainment, and other personal expenses, and refrain from paying those expenses out of your farm business account unless there is clear crossover. For example, if you are buying some work boots and coveralls, it may be a clothing expense, but it is clearly farm related, so you can use your farm account to pay for that. If a portion of your mortgage is paying for the land you farm (you can base this on the tax assessment, which probably assigned a value to the land separate from the improvements), you can write a check for part of your mortgage with the farm checking account and the portion that is the house you live in can come out of your personal account.

Getting all the financial management best practices implemented and running a profitable, sustainable family farm is not easy. It requires an intense dedication not really found in many other career paths. Even though the couple illustrated in the story below had other lucrative careers and probably a lot less stress carried home, Jackie and Harry Hoch felt the calling to return to family land and make a business of their own. Now in their sixth year of business, the Hoch family is balancing the management of many different crops, enterprises, marketing methods, and sustainability principles, while keeping their finger on the financial pulse of the business.

Strength in Numbers

HOCH ORCHARD & GARDENS, LA CRESCENT, MINNESOTA

• Jackie and Harry Hoch •

Not far from the Mississippi River, along the border of Minnesota and Wisconsin, sits an undulating piece of land growing over 70 varieties of fruits, berries, livestock, and poultry. Contoured rows of apples with names like 'Zestar' and

'Viking' curve along the hillsides, while a packing shed and commercial kitchen building sit atop the only flattish ground on the 94-acre property. A second generation of the Hoch (pronounced "Hoke") family occupies this land for now, scaling up what was once a hobby farm into a bonafide commercial business.

When Jackie and Harry Hoch first got married, they lived on Harry's family farm, trying to run a conventional wholesale apple orchard the best they could. The margins were just not enough to live on, especially once the barn needed rebuilding and Harry's siblings and mother's ownership stake had to be bought out. The couple knew they needed to earn incomes off the farm for some time to "save the family farm." They both continued their education and obtained good jobs in the Twin Cities, but the farm continued to occupy their future plans.

On weekends the couple would come back to the farm, steadily removing all the old, nonproductive apple trees that Harry's father had once planted and replacing them with high-density, dwarfing rootstock and more disease-resistant varieties. The Hochs also began to rebuild the infrastructure bit by bit. Now, 15 years after moving back to the farm, and just four years farming full time as a couple, the business is on solid footing, as long as the marketplace doesn't do anything drastic.

The couple's education has actually served them well in this new phase of the family farm: Harry has a bachelor's degree in integrated pest management and a master's degree in technical communications and sustainable agriculture; Jackie has a master's degree in business. The Hochs' two daughters, raised during most of their formative years on the farm, help out when they come to visit now. Although neither child may take over the family farm, Harry and Jackie would be just as happy to see one of their enthusiastic interns or perhaps grandchildren take an interest. They want to see the stewardship of the land and the business continue, but it doesn't have to be "owned" by the Hoch family.

The Hochs have created what they call a "fully integrated perennial cropping system," doing all of the propagation and production of the foods they sell (tree fruit, berries, vegetables, and animals). On the farm the family grades and packs all their fresh fruit, processing the stuff that's not of fresh quality into cider, jelly, applesauce, and other value-added products that they see a market for. They have recently added animals to use for flash-grazing the orchards, picking up fallen fruit, eating insects, and depositing fertilizer for the trees. Pigs, chickens, and ducks are moved about using portable electric fencing in a quick pulse through the orchard blocks. The animals are used more as a tool than another moneymaking venture, although the business has had no trouble selling the small quantity of meat they produce. When the animals aren't sanitizing the orchards, they are rotated around open pastures on the edges of the farm.

The Hochs move most of their fruit, berries, and value-added products through wholesale channels; namely, the plethora of food co-ops and natural foods stores in the Twin Cities region. They have found that consumers in the Twin Cities region are more willing to pay for premium organic fruit than people in areas a little

NICK MABE

Pigs flash-grazing through a young orchard.

closer to their farm, where local, conventional apples are being sold cheaply. Apples are the Hochs' main crop, with over 8,000 trees dotting their land comprised of 50 unique varieties. Hoch Orchard produces on the order of 150 tons of apples alone! With the addition of their early-season berries and value-added products such as honey, jellies, apple cider, and applesauce, Hoch Orchard can have a consistent year-round presence in the retail stores and keep the brand on people's mind.

The stores do a pretty good job of promoting the Hochs' fruit, sometimes labeling the different varieties with flavor profiles and hanging up pictures of the farm—one store even does a wall of Hoch apples in the fall. The Hochs make a point of inviting produce buyers out to the farm periodically for farm tours, creating awareness and farm loyalty that the stores can then pass onto their customers. A new, colorful, content-rich website also builds brand awareness.

Record-keeping and accounting are two areas of the Hochs' business that receive consistent attention. As certified organic producers, they maintain all of their organic system plans electronically and in paper files. They keep track of all agricultural inputs as well as animal rotations to stay in compliance: When their annual inspection occurs, Jackie can easily pull up any electronic file or receipt the inspector may need to see. Because of the on-site commercial kitchen and cidery, the Hochs have detailed food safety plans and practices that are kept in meticulous order.

Bookkeeping is done in levels. The first level of basic accounts receivable, accounts payable, and entering of receipts is done by a part-time employee supervised by Jackie. Jackie herself does the inventory tracking and invoicing. The second-level bookkeeping is done by a contract CPA, who generates quarterly reports, such as income statements, balance sheets, and cash flow projections. All financial records are kept in QuickBooks, with paper archives sitting in a nearby filing cabinet. Jackie pays particular attention to tracking accounts receivable so that their cash flow remains positive; the company snapshot function in QuickBooks; and calendaring the bigger payments, such as operation loan paybacks, into QuickBooks so she can anticipate potentially lean cash flow periods.

Although Harry and Jackie are not getting rich off farming (not financially per se), the family still manages to turn a profit each year, pay themselves a modest monthly salary, and cover their health insurance premiums. They also work very hard to provide living-wage jobs and educational internships to both locals and international students. The Hochs' commitment to being a socially responsible employer earned them one of the first Agricultural Justice Project (AJP) certifications in the country. With the addition

SILVANA CEDANO

Harvesting Hoch Orchards fruit

of more diverse crops, season-extension structures, livestock, and value-added products, the Hochs are endeavoring to offer more year-round employment, too.

When asked about how they financed the farm, they said that at first it was primarily off-farm income that allowed them to purchase the farm from their family and begin the early investments in new orchard blocks and infrastructure. Then, once the farm began making money and their equity increased, they could funnel that money into expanding and upgrading the packinghouse infrastructure and build cold storage. The couple was able to obtain a commercial loan to build the cidery and commercial kitchen because they had a proven track record and meticulous financial plans and had already established a market for cider. Prior to building their own facility, they spent four years purchasing a natural cider made off-site that included a blend of their apples and other Minnesota-grown apples, selling it to their co-op accounts and testing the market. Jackie and Harry then qualified for a highly competitive USDA Value-Added Producer Grant, which helped fund some of the startup operational costs of the cider processing and value-added production.

SILVANA CEDANO

Lower-grade apples get pressed for cider.

Each spring Harry and Jackie take out a modest-size operational loan from a local bank and pay it back in the fall after apple harvest. This alleviates the typical spring cash crunch that farmers face when they need to purchase a bunch of supplies for the season but don't have the money saved after the lean months of winter. Because they pay it back in seven or eight months, the interest charges are minor, and the predictable cash flow alleviates any money stress that Jackie might feel otherwise. Jackie always has a good handle on their cash flow projections, basing them on previous years' sales and an accurate product inventory tracking system to be able to predict future sales.

For example, if she looks back to November of the previous year to see that they sold 2,000 bushels of apples that month, then tracks that they have the same amount of apples in storage to sell in November of the current year, she can more or less predict the amount of apple income that will come in November and know how many expenses she can cover, too. If she had not kept good records of previous years or good records of current inventory, she would have no way to know how much apple income should arrive in a given month. Keeping good records also helps the couple keep track of their crop productivity and alerts them if there are issues in a particular orchard block or fruit variety.

Although the Hochs have not raised their prices in three years, they do review their pricing annually. They first look at their costs of production, then balance that with what the market will bear. To keep things simpler for the stores, instead of each of the 50 apple varieties having a different price, the Hochs have a three-tiered pricing system: regular varieties, premium varieties (those harder to grow or rarer), and club apple pricing. This makes the pricing easier for the stores and encourages them to carry more diversity without adding price complexity. Stores often display 8 to 12 different Hoch Orchard apple varieties at any one time, as well as their other fruits, such as cherries, apricots, and plums.

The Hochs have scaled up to be more commercially viable and to allow both of them to earn a living from the farm. They are now farming "in the middle," which is probably the hardest space for a farmer to operate in. They are too big to direct-market everything and too small to achieve some of the economies of scale that the thousand-acre apple farms of eastern Washington do. There are outside forces that they have no control over, such as expanding competition from cheap foreign apples, the rising cost of health care, and whether or not consumers will continue to value local and organic as their buying power continues to shrink. But the Hochs are committed to being responsible land stewards, good employers, and producers of high-quality, healthy food, what they call the "anti-Walmart." It remains to be seen if our culture of cheap food will reward the Hochs for their values.

TAXES

I am not going to discuss taxes too much because the regulations change every single year. In general sole proprietorship farming operations should use a Schedule F tax form or a Schedule C Small Business tax form. For farms that are more than just farm production, such as more vertically integrated operations or those that do value-added products, a Schedule C might make more sense for you because it includes a lot of business tax deductions that the Schedule F does not include. The Schedule F is designed for basic commodity crops and those receiving various government payments. If your farming business is incorporated as an LLC, S-Corp, or some other entity, you should find out the most appropriate form to use from your CPA or from the IRS website. My only advice for taxes is:

- If you are thinking of evolving from being a "hobby" farmer to being a commercial farmer, you may want to take a peek at the Schedule F instruction book and consult with a tax professional who specializes in agriculture before starting a legitimate business.
- Get it done on time. Businesses often have to submit their taxes even earlier than personal income tax forms—check with the IRS for your due date. You can always apply for an extension if you need more time, though.
- Be thorough, making sure you include all of your farm-related income and farm-related expenses.
- Be careful about trying too hard to reduce your current year's tax liability. If you take every deduction and all your allowable asset depreciation now so that you don't end up owing any taxes, not only might it reduce your options in future tax years (such as taking depreciation), it might also harm your ability to qualify for a loan in the future because your profitability will look poor. Showing a loss every year may earn you a nice tax return check, but it won't earn you any points with potential lenders, investors, or partners or when you want to sell the business.

ENTERPRISE BUDGETS

Not only should you create a whole-farm cash flow budget, you should make budgets for each one of your enterprises, called **enterprise**

budgets. These are estimates of the income and expenses required to produce a particular crop, animal, or other farm-based enterprise. Many university Extension offices produce enterprise budgets for typical crops grown in your state, but use them only as a guide. I draw on them to remind myself of some of the expense line items that I otherwise might forget about. Richard Wiswall's book, *The Organic Farmer's Business Handbook,* goes over enterprise budgets in detail and contains good advice about picking the best mix of enterprises to maximize profitability. Even as a diversified crop producer, Richard looks at each crop individually, using an enterprise budget to determine the right mix for his farm and financial goals. In doing this he finds out details such as that kale is more profitable than broccoli because you can harvest much more off a single kale plant than you can off one broccoli plant.

Whether you are just starting out or you have been growing the same three things for three generations, you should run through all your enterprise budgets. In those budgets make sure you include all the costs of production, including the **variable costs** (those that change with the amount of production, such as seed and feed) and the **fixed costs** (those that do not change based on production, such as some insurances, loan interest, capital assets, property taxes, advertising, asset depreciation—things that are often referred to as "overhead"). Once you know the net income for each enterprise, you should then compare each enterprise based on labor hours to find out the *net returns per hour of labor* invested. This may help you to really focus on those things that are the best "bang for your buck"—basically, the most profitable for your investment of labor. Here is a little table that outlines how you do that (all enterprise numbers in this table are fake):

ENTERPRISE	NET PROFITS PER UNIT (BEDS, ACRES, ANIMAL UNITS, ETC.)	HOURS OF LABOR PER UNIT	PROFITS PER HOUR
Hogs	$250.00/head	2.5	$100/hr.
Broiler chickens	$2.75/head	0.5	$5.50/hr.
Hay	$2.50/bale	0.057	$43.86/hr.
Watermelons	$195/acre	7.35	$26.53/hr.
Potatoes	$75/acre	6.4	$11.72/hr.

Of course, a lot of factors have to be considered when you make decisions, such as your holistic goal or your market demand. However, from this table you can see that your labor is best used for the hogs and hay, with watermelons in third place. You would especially want to reconsider whether broiler chickens were an essential enterprise for you, given the low returns and higher labor requirements. (Again, this is just hypothetical and not real numbers, although we found the same thing when we analyzed our broiler production compared to other enterprises, which caused us to drop broilers after three years.)

TAKE-HOME MESSAGES

- Set up good bookkeeping practices from the beginning. You don't have to use QuickBooks or some other computer program right away, but at the very least begin using a paper ledger and organizing your receipts.
- Save all your receipts, including the things you pay cash for. If you perpetually lose cash receipts, consider paying by check or debit card. I eventually had to tell my husband to stop paying with cash and that he had to use the debit card for everything because at least I had bank records of those transactions.
- Keep a notebook, record sheet, or some other form you create to keep track of cash sales on a regular basis.
- At least once a quarter generate a profit and loss (P&L) report to keep track of where you are. When you get more proficient, try to do this monthly. Don't get too far behind, or you may not notice right away when your expenses are more than your income. Aim to have all your transactions inputted by the tenth of the following month so you can generate an accurate report for that previous month.
- Start noticing seasonal cash flow crunches, and try to prepare for them before they arrive. For example, prepay the feed mill when you have money to cover your feed requirements when sales slow down but you still have mouths to feed. I know farmers who even buy themselves gift cards from their favorite seed or farm supply store when they have good cash flow, then turn around and spend that card in the late winter or early spring when they need to buy supplies the following season. Do whatever creative arrangement you need to do.
- If you want to be a sustainable farmer, you need to run your farm like a business (albeit a socially responsible one). Focus on cost containment (get creative on ways to reduce certain expenses), pricing your products properly, and selling the volume at the prices you need to cover your costs and make some profit.

- You can't just pass all of your increasing expenses along to your customers without making at least some effort to improve efficiencies and cut costs. For example, say you regularly have a 10% death loss in your poultry and you pass that cost onto your customers to cover for the lost birds. Are you simultaneously working on methods to reduce your death loss? If you could get your death loss down to 5%, could you also slightly reduce the price of your product? Your customers should not have to pay for your inexperience, ignorance, laziness, inefficiencies, poor labor management skills, and so on. Try to get better at those things without compromising your values, product quality, or quality of life.
- Find out what a typical enterprise budget looks like for the crops and animals you produce. If your income or expenses look dramatically different from those budgets, do some research to find out why they vary so much. It may be that some of your expenses are well beyond the norm and can help point out areas for you to do more investigation into operational efficiencies and cost containment. A rather extreme example, but sadly not that unusual for what we saw in our visits with farmers around the country, was a pastured-poultry farmer we met who spends two hours a day filling up waterers for his large laying hen flock. If he looked up a free-range egg enterprise budget on the Internet, he would not find any labor costs associated with watering the flock, because the budget assumes the farmer has an automatic watering system. If this farmer had researched typical poultry enterprise budgets, he might decide at once to invest in automatic waterers.

  However, oblivious to this information, he will continue to spend around $11,000 a year on labor costs for watering (his employee is paid $15 per hour). An automatic system would be just a fraction of that cost and last him for many years to come. This particular farmer is not currently profitable and is always on the verge of throwing in the towel. He could be putting $11,000 in his pocket each year, which might turn his business around for the good.
- High-quality bookkeeping takes time, time that you may not have. Instead of ignoring the bookkeeping, look for someone who might help you. You can hire a regular bookkeeper or perhaps you have a customer who will do your bookkeeping for free or in exchange for food. Make it a priority early on in your business to keep good financial records. This will reward you in many ways in the future.

- Take on debt and make capital purchases wisely. Select debt that will be productive for your business, not just immediately depreciate. This may include debt to hire a specialized employee/manager who can help take your business to the next level. An example of productive debt might be a team of oxen and the related implements. They provide labor, fertilizer, and perhaps meat (not to mention enjoyment, customer interest, and so on) An example of nonproductive debt might be re-siding your barn so it looks prettier or buying an expensive new implement that you use only once a year for primary tillage.
- Fill out the proper tax forms, even in your first few years of farming. Even if you only sell $5,000 in product, you should start filling out Schedule F or Schedule C tax forms right away. Establishing this record will allow you to apply for loans in the future. Use Schedule F if you have a fairly simple or traditional farming operation and are not vertically integrated. If you run a more complex, diversified farm, or one in which you have vertical integration or add-on nonfarming enterprises such as classes, you may want to use Schedule C (small business tax form).

CHAPTER 12

HUMAN RESOURCES AND FAMILY

THE CONCEPT OF SUSTAINABILITY also includes a social component: How do you treat all the stakeholders in your business? Stakeholders include the owners (counting you!) and their families, employees and their families, interns or volunteers, neighbors of the farm, other members of your community, and your customers. You can draw the circle much wider than that if you'd like. If your business is built upon exploiting any of those stakeholders, you will not be sustainable over time (although you may be financially "successful"). And why bother anyway? There are plenty of businesses built on the backs of people—why would you want to emulate that practice by becoming yet another one?

Although food producers are often squeezed on both sides by the increasing cost of doing business and the lower prices they receive for their efforts (known as the "cost-price squeeze"), there are many ways to improve the way you treat the human resources of your operation that won't necessarily squeeze you even harder. In fact, they are likely to improve your business longevity over time and create a happier workplace for all, yourself included. This book is about creating a new paradigm for doing business, one that will last into the future, taking care of both our planet and all of its inhabitants (human or otherwise). This chapter will include ways to be good to all of your stakeholders and keep everybody as content as possible, while creating a business model that lasts. We will also talk about bringing on employees and what all that entails. Let's start with the immediate farming family.

FAMILY-FARMING PRINCIPLES

What is a "family farm"? The USDA Economic Research Service defines one as "a farm in which ownership and control of the farm business is held by a family of individuals related by blood, marriage, or adoption. Family ties can and often do extend across households and generations." You could be a family farm as a single person going it alone, a couple or group of siblings managing it together, or a multigenerational farm with parents, children, even grandchildren that have a say in the management. The exact arrangement or players in a family farm really don't matter—there are best practices that pertain to all of them. Here are some best practices that we found in our nationwide conversations:

Successful family farms . . .

- **Make goal-setting a priority.** They work to form a holistic goal together, as described in chapter 5.
- **Communicate well, and communicate often.** They make time for family meetings and work to keep personal issues and drama out of them.
- **Make time for things other than farming.** They try to find balance, time for fun, personal health and wellness, and each other.
- **Place people in roles that they are good at and want to do.** If a nonfamily employee is a better fit for a job, then that person is placed there. Family members don't get a free pass and don't get to assume business positions they don't have the skill sets for. Likewise, family members are not used as free or convenient labor just because they are around.

Although immediate family members do not *have* to be paid in most family farms (unless you are incorporated), they *can* be paid and their wages written off as a business expense. Certain payroll taxes are not required for immediate family members, either; make sure you look at the latest federal and state regulations regarding family wage laws. Even if they are not paid for their contribution to the business, family members should be covered under the business's workers' compensation policy. You don't want them getting hurt on the job and not having any coverage. Family members also can be compensated with an ownership stake, receive profit sharing, or have their living expenses covered under the "owner's draw" if they live in the same dwelling as the owner. State laws vary, but extended family members (cousins, uncles, and so on) will probably have to be paid wages, just as any other employee will. More on the legalities and tax implications of employees later in this chapter.

Children on family farms often do work on the farm or in other aspects of the business, such as helping with marketing or doing some basic bookkeeping. Nowadays, many teenage children might be better equipped to handle the business website, social media marketing, and other computer work because of their immersion in these things, often beyond what their parents have experience with. Children can derive a lot of benefits from working in a family business and outdoors on a farm, from building skills and confidence to building physical strength to helping them home in on their career goals and have something to put on their resume. I don't need to remind you that you are a responsible parent and won't exploit your children, but apparently, it does happen sometimes on family farms.

Instead, work to create a positive work environment for your children, such as including them in farm business meetings, goal setting, employee training, safety training, and everything else that your other employees and family members participate in. Stress safety above all else, and be mindful of their competing needs for schoolwork, socializing, nonfarming hobbies, rest and relaxation, and so on. Also, pay attention to state and federal laws governing child labor on family farms. The laws are always changing; make sure you know what they are before you ask your child to become involved in the farm work. Although you don't have to pay your children, you should probably come up with some sort of fair system of remuneration to honor their contribution and keep them motivated. Maybe working four years on the farm part time gets them four years of free college education that you fund? If you want your kids to take over the farm business some day, make sure to involve them early on and give them a positive experience. Otherwise, as I sadly see too often, the farm children will run toward the city and more urbane careers, fleeing from a career in agriculture that looks either too hard, miserable, or poor.

HUMAN RESOURCES MANAGEMENT

You may need more than just yourself or your family members to grow your business or simply get everything done that you want to in your busy days. We realized this ourselves about three years into our farming business. We first started by hiring some college students as part-time employees to work our farmers' market booth. This gave my husband more time for farm projects on weekends and gave me time to play with our daughter since I already held down another full-time job. Sometimes you have to bring on employees just to have some balance in your personal life, but often, it is worth every penny. We then hired a part-time employee to help butcher chickens with us two times a week for a couple of years, and when we dropped the broiler production enterprise, we switched this employee's focus to our laying hens operation. She moved to working half days washing and packing eggs.

We eventually hired a couple of young men to do the daily animal chores (feeding, watering, egg collecting and packing) while my husband focused on building projects, equipment maintenance and repair, meat processing logistics, and capturing waste food streams for our animals. I explain all of this because I want to elucidate a few good reasons for hiring help. If you have basic, daily chores that can be taught to someone else to perform, maybe you should delegate that role to an employee. Sounds obvious, but often, it is not. Many business owners, farmers included, think only *they* have the skills and attention to detail to run their entire operation and are not willing to let go of control and delegate some tasks to others.

Perhaps you don't think you have the income to justify hiring employees. As a farm business owner-operator, you have to wear a lot of hats, including business planner, equipment repair person, tiller and planter, animal husbandman, harvest and postharvest manager, accountant, tax expert, legal specialist, human resources manager, public relations person, marketer extraordinaire, and salesperson. If you can delegate some of those tasks to the right person or team of people, it will often allow you to do a better job with the other hats that you wear and maybe make the business more profitable and sustainable, or improve quality. I have heard this referred to as working "on the business" instead of "in the business." It may not be the direction you hope to go—that is, spending less time in the fields or at the markets—but it is a typical trajectory for a growing and maturing business. The owners eventually spend less and less time on the "operations/production" side of things.

Otherwise, you can choose to keep things smaller and simpler so that you and your family can do it all without employees. Many farmers we've met have purposely chosen a small-scale business model so they won't have to deal with employees. But having a better understanding of human resource management, how to engage employees, and how to create a positive work environment may equip you with the skills to be able to successfully hire and retain a quality workforce that helps improve your business viability.

Human resource management is a process that can be broken down into specific activities: job analysis; writing job descriptions; hiring, orientation, training, and professional development; employer/employee interactions; employee engagement; performance appraisal; compensation; and conflict resolution/discipline. I will go into each of these activities in a little more detail, with specific advice for farm employers.

Job Analysis

Too many farmers wait until the need for employees is at a critical stage before they start their search. When the pruning needs to happen right now, or the weeds are getting out of control, or the sheep will begin lambing any minute, you have waited too long to find extra help. This is a serious issue that is not taken all too seriously by many farmers, especially those that are still in the inexperienced startup phase and may not understand their labor needs. Develop a human resources plan to identify areas where additional employees may be needed (what time of the year, how many hours, what specific activities), and do an annual review of this plan to keep it up to date. Perhaps you want to break it down by season (winter, spring, summer, fall), by month of the year, or by major activity (e.g, pruning, planting, spraying, irrigation, staking/trellising, thinning, harvest). Figuring out what jobs you need covered will lead you to write accurate job descriptions.

Job Description

This is all about the what, where, when, and how of the positions you plan to recruit for. The more detailed you are, the more likely you will find candidates that understand your expectations and what their roles would be. However, if you expect the job to vary somewhat over time, make sure you include some of that flexibility in the job description. For example, if you want to start the position as part time but expect to have it grow into full time by the year's end, mention that in the job description. I think it's important to list the compensation, but do so as a range so you can adjust what you offer based on the person's experience and skills. For example, you can write:

"Hourly wage of $11–15/hr. DOE [depending on experience]. Additional compensation *may* include on-site housing, farm-produced food, transportation subsidy, and flex-time benefits."

Hiring

Finding good help can be challenging, and I don't have any easy answers for what proves to be a nationwide problem finding competent employees in agriculture. I don't think it is a wise business move to rely on an illegal workforce, so I'm not going to suggest you go that route in your hiring because it could be placing your business in serious legal risk. Here are a few places you might look for skilled (or at least enthusiastic) legal agricultural employees:

- Local high schools, especially agriculture students (Future Farmers of America [FFA], 4H) or perhaps business students to help with writing plans or marketing (such as the DECA club). Be aware there are different labor laws that apply to youth 16 and under, and they are constantly changing.
- Local community colleges and universities
- Naturalized citizens or green card holders (permanent residents) with backgrounds in agriculture
- Unemployed 20- or 30-somethings (at least their backs and enthusiasm tend to be strong!)
- Returning war veterans or discharged soldiers (crazy work ethics!)
- Underemployed culinary professionals (they know how to cook, great for marketing positions!)
- Wannabe/aspiring farmers
- Unemployed MBAs (especially for everything marketing and sales related, as well as financial management)

Write up a detailed job description as mentioned in the previous section, distribute it online and in print around your region, interview the best applicants, and call their references. As part of the interview process, give them a thorough tour of your farming operations, especially the areas in which they would be working. If feasible, show the prospective employees the specific activities they would be doing, such as milking cows, collecting eggs, planting transplants, washing and packaging product, or selling at the farmers' market. Make abundantly clear the range of environmental conditions they would be working in, any productivity expectations you may have, and whatever rules of the farm you maintain. The more you lay out up front, the more likely you will reduce conflict later on.

Orientation

Once you have selected the best employee for the position, make her a job offer in writing and orally. Give her a chance to ask any questions she might have. Once you agree to a start date, spend the first week of work in orientation mode. What does orientation look like? Even if the employee already received a farm tour during the interview process, do it again. Cover things in more detail, such as parking locations, emergency procedures and exit routes, farm rules, break room/rest areas, and food safety protocols, and introduce her to other members of the farm team.

Create some sort of employee handbook, and have employees read and sign and date it. If an employee does not speak English, try to translate the handbook into his native language and explain it verbally to him in his native language. Since an employee handbook has legal ramifications, it should be made understandable to everyone you hire. The employee handbook should include your house rules, the specific job description, legal requirements (wages, overtime, and so on), grievance procedures, food safety protocol, and a safety training section. If this is scaring you, it doesn't have to. I have seen employee handbooks as short as five pages, and you can often find templates online. However, the more you include in the handbook, the more protections you are creating for yourself and the greater clarity you are providing for your new employees.

Let new employees know when and how they will be paid and explain how all of your other compensation/benefits programs work. Have them fill out any necessary employment forms, such as I-9s, W-4s, and insurance enrollment forms. Go over the method you use to record hours and any other recordkeeping forms that you expect them to fill out (such as production records, harvest records, temperatures, and hazard analysis and critical control points [HAACP] forms).

Training and Professional Development

You may want to create what you call a "probation" period for the first three to six months in which employees are paid less or receive fewer benefits while they are in training. This gives you a chance to "test" them out before fully committing to offering long-term employment and your full compensation package. If you are only hiring temporary, seasonal employees, then this concept may be moot. Or you could call their first week their "probation" time. During the probation period, watch their pace, attention to detail and quality, and ability to hear and execute the tasks you give them. If someone lacks particular skills but demonstrates an eagerness and capacity to learn, she just may work out well for you. Give her space to ask questions and gain additional training if need be. And be a stickler about safety during the new employment period—when someone is inexperienced with a task, an injury is much more likely. Also, it is your job as a good employer to give feedback during the probation period. Your employees cannot read your mind and can't improve if you don't communicate frequently with them. Building feedback loops sets a culture of respect and continuous improvement.

Remember the following statistics when you are in the training phase. Your new employees will be much more likely to remember and understand a task if they are given a chance to perform it for themselves, with your supervision.

A learner tends to remember:

1. 20% of what he hears (your verbal instructions)
2. 30% of what he sees (while he watches the task demonstrated)
3. 50% of what he sees and hears (having the task explained and demonstrated at the same time)
4. 70% of what he says (verbally repeating steps of the task)
5. **90% of what he says and does** (telling you the steps while performing the task himself)

In addition to on-the-job training, you should consider other professional development opportunities for your employees that go beyond just their job description. You can facilitate employees' acquisition of new skills and areas of knowledge through informal and formal training opportunities, such as mentoring and coaching, sending them to workshops or conferences, or enrolling them in a formal certificate program. Several farmers I know will send their employees to an annual sustainable farming conference, where they learn about a diversity of farm- and business-related topics, meet interesting people, and often get inspired. It is not a huge cost for the farm because there are often scholarships for lower-income people to attend conferences—the farmers just pay for the day's wages of their employees and facilitate the group transportation.

But the impacts of a day or two at a conference on your employees can be long-lasting. Providing professional development opportunities can help retain good employees and provide them the means to take on more responsibilities and specialized tasks. It is usually a win-win decision on your part to offer professional development to your staff.

Employer/Employee Interactions

How do you retain good help and keep them motivated to do their best work? Apart from fair wages and benefits that increase as a worker's productivity and business profits increase, studies say that what employees most often want is **respect**. Respect can be shown in a myriad of ways, but most important is providing your employees the time and space to share their concerns, ideas, and opinions and to be involved in decision making. The manner and tone in which you communicate is also important. Talk to people as though they are adults, and say "please" and "thank you," even when you are telling folks what to do for the day. Meet as a team at the beginning of the shift, mention all that you hope to be done in that day, and ask for feedback on how it should all get done. Stress safety above all else, which shows you respect their health, and stress quality, which constantly reinforces the mission of your business. Speaking of mission, to the degree possible, involve your employees in your annual update of the farm and business's holistic goals. That will further reinforce that they are part of your "team."

Communication is key, as I have mentioned several times already. But how can you do it effectively? One of my favorite communication methods is called nonviolent communication (NVC). We don't usually think we are committing violence through our words, but we often and actually do. Verbal abuse can induce fear, stress, anger, despair—a whole range of deleterious mental states. So how do you communicate without using violence? You could spend years in NVC training and still not have this system memorized. I have taken over 60 hours of training and am still only marginally successful with it. However, I use parts of NVC all the time and adapt it to different situations. Here are the four basic steps of nonviolent communication:

- **Observation**—"When I see you do this (be specific)________, I feel this________."

- **Need**—"Because my need for _______ is not being met."
- **Listen to them and give empathy.**
- **Solution—**"Would you be willing to do _______?"

Here is an example of how you might use these four steps in a conversation with a challenging employee: "*Todd, when I observed you talking on your cell phone while on the picking ladder this morning, I felt a bit scared for your safety. I really want to create a safe work environment and don't want anybody getting hurt. Would you be willing to leave your phone in your car during work hours? You could still check it during your breaks. How does that sound?*"

You can also use steps 1 and 2 to give compliments, except talk about your positive feelings and your needs being met in some way by the actions of the other person. It makes for a much deeper and meaningful compliment when you put it in the context of how it makes you feel. Here's an example: "*When you stayed late the other night while the sheep were lambing, making sure all the ewes and newborn lambs were doing okay, I felt extremely grateful to have found an employee that cares so much for our animals' welfare and our need to get some sleep after this last crazy week of lambing.*" Make sure you remember to give compliments to your employees, not just criticisms.

When talking with an employee about performance or behavior issues, try to be as specific as possible about an incident, instead of casting a label on them, such as "*You are lazy*" or "*You don't pay attention.*" When you use generalizations like that, it usually makes the other person defensive and unable to hear you.

Try to keep your voice calm. If you expect conflict, perhaps bring in a more neutral third person to facilitate. The facilitator can repeat back what you both say, but in a calmer, clearer tone that can diffuse anger.

Employee Engagement

Engaged employees find satisfaction in their work and develop feelings of commitment to the business. Engaged employees also have lower rates of absenteeism, turnover, shrinkage (theft), safety incidents, and higher rates of customer advocacy, productivity, and profitability. But how do you engage employees, particularly in a farm business setting? (I have adapted the following points from a Rutgers University paper, "Employee Engagement: Strategies for Creating a Positive Work Environment.")

- Show them the importance of their jobs to the success of the farm business.
- Align employee performance goals and employee development (training) to personal and business goals.
- Help employees understand how to succeed in their roles and tasks. Don't assume they know what you want—they are not mind readers.
- Communicate effectively through regular feedback, coaching, and recognition. Learn how to give compliments and thanks.
- Improve work processes to enhance employee participation (through regular staff meetings, team decision making, and so on).
- Show employees how they can develop through their work; foster an empowering culture of respect and trust.
- Encourage autonomy, participation, and considered risk. This is a hard one, but you have to learn to let go of some control.
- Have regular meetings and provide employees a chance to share ideas for improvements.
- Provide trainings and good delegation so your employees will become competent decision makers and can function with less and less supervision. Empower them to shine. Cross-training is also important so they possess a wider breadth of skills and can help out when another employee is sick or leaves.
- Share basic company financials such as profit and loss statements, sales goals, and budget versus actual with your team so they have a better understanding of the business and how important their respective roles are.

Performance Appraisal

Apart from the daily and weekly informal feedback you give your employees, take some time to do an annual or semiannual performance review with each of your employees. Create a template that you use for all employees and sit down, face to face, with the employee as you talk through all of the questions. Try not to focus just on the employee's weaknesses but also discuss his strengths. Give space for his feedback on how *you* are doing as a manager/supervisor and any ideas he may have for fostering a more positive workplace. Also include a question about any desires he has for further training and other ideas he might have for improvements. If there are areas for improvement, make sure you follow up on your commitments and the progress of the employee; otherwise, it makes the performance appraisal appear to be a useless exercise.

Compensation

Pay what you can afford and what you think the employee is worth, but pay at least minimum wage, even if the person is an intern. Interns have to be paid minimum wage, or their combination of a stipend and room and board must equate to at least minimum wage in most states. Otherwise, you open yourself up to potential legal issues, in addition to which you may be exploiting that person. More on this issue later.

Consider building opportunities for performance or merit raises or bonuses into the wage structure. If you start an employee at $14 an hour, you may not be giving yourself any space to offer a raise. Consequently, that person may be less motivated to perform better or go above and beyond the basic job duties because she knows she can't get paid more. Maybe you can offer profit sharing, performance bonuses, or pay plus commission to incentivize excellence, but start with a lower wage. Profit sharing is also a good "insurance" mechanism that allows you to better compensate your employees during profitable years but does not increase your labor costs during bad years.

In addition to wages there is a plethora of compensation and employee benefits options that lead to a positive, healthy, and productive workplace. Here are some ideas:

- **Offer health benefits**, such as employer-sponsored health insurance, contributions to health savings accounts, or paying for yearly preventive care. You could also bring the doctors to your employees (especially helpful if you have a fair number of employees) through annual blood pressure checks, cholesterol screenings, and even dental cleanings. You can even invite a masseuse or chiropractor out for group treatments (who doesn't like a good back rub?).

  Other innovative ways to improve employee health and wellness is giving them farm-fresh food to take home, access to personal gardening space, free bicycles, and even gym memberships. Some employers pay for smoking cessation courses or help their employees commit to drug or alcohol counseling. If you don't have a lot of financial resources to pay for costly health insurance and benefits, be creative—there are other ways to improve your employees' health and wellness while preventing injury, illness, and even disease (and for their families, too!). You can make sure your employees know where all the low-cost clinics are, help eligible employees sign up for food stamps or Medicaid, and provide paid time off for sick days or at least some flexible hours so they can see the doctor when needed. As a farmer you are uniquely situated to make one of the biggest contributions toward your employees' health by improving their diets and food security!

- **Ease transportation needs**, through practices such as free bicycles for workplace commuting, helping your employees arrange busing or carpool, or maybe give them a low- or no-interest loan for a safe, fuel-efficient scooter, motorcycle, or car. Incentivize green transport through buying bus passes, providing free bikes, installing showers for bikers, and offering raffle prizes to regular carpoolers, bus riders, bikers, or walkers. Assist them in finding housing closer to the farm or perhaps even provide on-farm housing so they don't have to commute as far. If you provide transportation to work with a bus or van, for example, make sure it is safe and up to code with things like seatbelts.
- **Be a stickler about safety** through regular trainings, safety gear, ergonomic trainings and equipment, allowing sufficient time to complete tasks, and so on. Also, don't let people drink or do drugs on the job. We have heard way too many sad tales of serious injury and even death on farms, particularly with younger and less experienced farm employees, due to careless behavior.

  While safety training is not a form of compensation (every farm should do it!), going above the norm by providing free personal protective gear, ergonomically approved tools and equipment, and opportunities for advanced training can be a benefit of working for you. I know some farms that give all their employees advanced first-aid training: This not only creates a safer workplace, but it is also a valuable skill for your employees and something they can list on their résumés.
- Provide some **paid vacation and sick time** if possible. You don't want employees coming to work sick, but they will if you don't give them paid time off for staying home and recuperating. This is also a food safety issue: Your employees should not be coming to work sick and spreading germs. Let them stay home until they are no longer contagious.

  If you can't pay vacation leave, maybe you can take them on a little vacation. Full Belly Farm, a large organic farm in California with 50 or so employees and apprentices (I was one of them), paid for the entire crew to go whitewater rafting for a day. It wasn't a full-fledged two-week paid vacation, but it was an enormous morale boost, and we all got to know each other better. An Oregon farmer I met recently buys round-trip plane tickets for all his employees to go home to Mexico for a month at Christmas! He can't pay for that month of work, but they do get free plane tickets and a guaranteed job when they get back. This yearly treat has created tremendous loyalty for him—all but one of his employees has been with him for over 12 years!

Conflict Resolution, Discipline, and Dismissal

If employees are making mistakes, not upholding quality, working too slowly, or any other things that are frustrating you, address it immediately before letting the problem linger or amplify itself. Talk to the person; let her know your concerns and give her a timeline to correct the issue. If this doesn't work, you can either give her another chance or dismiss her. If you do give another chance, let the employee know what the consequences are for not making the changes, and be firm about this. If you let any employees get away with poor performance regularly, you are potentially jeopardizing your business, and you will lose the respect of your other employees.

A little personal example of this was when a part-time employee of ours was seen taking short breaks all day long to text-message his girlfriend. Our other employee would see this and get annoyed because he himself was busting his butt working steadily all day except for a lunch break, while the first employee was taking texting breaks throughout the day. The "texting addict" employee was not dealt with, and his work got shoddier; likewise, he didn't get all his daily chores done. We mistakenly let the problem fester and just stopped calling him to come into work. The texting addict probably never learned about his performance issues, and the other employee lost a little respect for us. We should have communicated the problem to this employee early on, and perhaps it might have stopped the texting issue in its tracks.

After this experience, though, we were a little hesitant to hire young twenty-somethings whose fingers seem glued to one electronic device or another. Don't let one bad experience cause you to label or stereotype a whole group of people. I think the clearer you are about job responsibilities and farm rules early on, the more you can avoid conflict later. Maybe one of our farm rules should have been: Cell phones get left in employees' cars!

If you are having conflict in the workplace, either between employees or between an employee and you (the owner), consider utilizing some sort of conflict resolution process before the situation spirals out of control or leads to employee dismissal. As you are probably aware, conflict in the workplace can be incredibly destructive to good teamwork, productivity, and, ultimately, your farm's viability. Managed in the wrong way, real and legitimate differences between people can quickly spiral out of control, resulting in situations where cooperation breaks down and the team's mission is threatened. This is particularly the case when the wrong approaches to conflict resolution are used, such as anger, guilt trips, or passive-aggressive communications. To calm these situations down, it helps to take a positive approach to conflict resolution, in which discussion is courteous and nonconfrontational and the focus is on issues rather than on individuals. Use the nonviolent communication methods I described above. If this is done, then, as long as people listen carefully and explore facts, issues, and possible solutions properly, conflict can often be resolved effectively.

What happens if you do have to let somebody go? Keep notes or records of how he underperformed or didn't complete his job responsibilities. Then give the employee a full explanation (preferably in person, as well as in writing), let him explain himself if he so desires, then give him a set period of time to pack up his things, collect his last paycheck, and go. Try to keep things professional and cordial to prevent any extra animosity and avoid legal trouble. If you are concerned about a potential blowup or unsafe response, or even a legal challenge, try to have another member of your ownership team present as a witness at the dismissal meeting.

APPRENTICES, INTERNS, AND VOLUNTEERS, OH MY!

It can be a fantastic experience to have apprentices, interns, or volunteers working on the farm. You may learn things from them as they simultaneously learn from you. You get to perfect your skills as a mentor, an educator, and even a communicator. You may form a lasting friendship; you may bring some liveliness and even companionship to an otherwise fairly quiet and sometimes solitary life running a farm. But if you host interns, apprentices, or volunteers, make sure you know the laws in your state and at the federal level; otherwise, you could get yourself into trouble, even when you think you are doing good. Recently, many small farms have been fined thousands of dollars by their state's department of labor for not following the law when it comes to interns and apprentices. Here are some of the federal definitions and their implications for your farming business so perhaps you can avoid trouble in the future:

Interns

An internship is a work-related learning experience for the benefit of individuals who wish to develop hands-on work experience in a certain occupational field. Under certain conditions individuals may—without any expressed or implied compensation agreement—work for their own advantage on the premises of another and are not necessarily employees. Whether they are considered trainees or employees depends on all the circumstances surrounding their activities on the employer's premises.

The most likely way an intern may be considered a nonemployee and exempt you from wage and hour requirements is if that person meets the **six-factor test** as set forth by the US Department of Labor (DOL) for "trainee" or intern status. According to the DOL, each of the following six criteria must be met:

- The training, even though it includes actual operation of the facilities of the employer, is similar to that which would be given in a vocational school.
- The training is for the benefit of the trainees.
- The trainees do not displace regular employees but work under close observation.
- The employer that provides the training derives no immediate advantage from the activities of the trainees, and on occasion his operations may actually be impeded.
- The trainees are not necessarily entitled to a job at the completion of the training period.
- The employer and the trainees understand that the trainees are not entitled to wages for the time spent in training.

Some states require that interns be registered students at an accredited institution and receive academic credit for their internship. Contact your state's department of labor (or labor and industry, as some are called) for the latest rules. If all of this scares you, just get your interns on payroll and on your workers' comp policy and pay them at least your state's minimum wage, and you won't have to worry about complying with the six-factor test above. However, this can be costly for you, especially if the intern is a newbie and lacks the skills of a paid employee. Not many businesses can justify paying regular wages to somebody in training.

Apprentices

An apprenticeship has to be registered with the state or feds to be considered legal. You have to pay at least minimum wage (unless the apprentice is under 18 years old), and you are expected to pay more over time as the apprentice's skills improve. The concept of apprenticeships is that the person will eventually be skilled enough to have the occupation—in the case of farmers, the trained apprentice would be equipped to run her own farm. Draft an educational curriculum lasting at least 144 hours, demonstrate that you are a qualified instructor, then register with the federal government. Some states are piloting approved farm apprenticeship programs; Oregon and Washington are two of the most recent. Check with your state department of labor to find out the latest. If you don't want to go through all of this rigamarole, don't call the position an "apprenticeship."

Volunteers

Some organizations use volunteers, and there has been some discussion on whether farms and growers could use this classification for their interns. The ability to use volunteers is very limited, if not impossible, in a **for-profit** organization. Legally, the following are criteria for volunteers and employers who wish an exemption:

- Volunteers must perform their services for civic, charitable, or humanitarian purposes.
- They must serve without promise, expectation, or receipt of compensation, except for expenses, reasonable benefits, or a nominal fee that is not tied to productivity.
- They must provide services freely and without pressure or coercion from any employer and offer their services solely for personal reasons or pleasure.

In all cases (interns, apprentices, and volunteers), make sure you provide them opportunities for skill building and training in addition to varied farm labor experiences. Maybe take them out to tour other farming operations in your region, and allow them to meet other interns, too. In my best internship experiences, I was welcomed in as almost a member of the family—and I learned so much more as a result. Intense and lively discussions over the dinner table gave me a deeper understanding of the business side of farming, challenges in the marketplaces, and a good dose of what I call "farm politics"—the things happening beyond the farm gate that were constraining the farmers' ability to do their job.

As I mentioned above, your state will probably require that you pay your interns and apprentices minimum wage and withhold payroll taxes, unless you can follow all the rules of "vocational training" and "approved curriculum." If you choose to offer a paid stipend along with room and board, make sure that the sum of those benefits equals at least minimum wage. If you have any questions or concerns about this issue, you might want to put in an anonymous phone call to your state's department of labor. In addition to wages draw up a little contract that spells out the intern/apprentice responsibilities along with what kind of training you will be providing them, as well as any form of compensation they will receive. This will clarify things and can prevent future conflict by "managing expectations."

Even if your intern/apprentice is new to agriculture, listen to his ideas and give him small responsibilities over time that will build his skill set and his self-confidence. You might even provide him with the resources to undertake his own project, whether it be planting and caring for one bed of vegetables in your field, building a new chicken coop, or selling at a new farmers' market. If you hold onto all the power and control, your intern/apprentice might become dissatisfied or lose some of his eagerness and enthusiasm. Empower him to improve his skill sets, and he will likely shine.

PAYROLL

Payroll taxes are fairly complicated and vary state by state. I suggest taking a class: You can often find one at your nearest department of labor, department of economic development, or perhaps even your local chamber of commerce. Even if you choose to pay an accountant or payroll service to do the job for you, you should still know the laws. In my opinion, if you have more than two employees, you should consider contracting out your payroll services, which includes filling out all the reports that go to your state and federal tax boards. This will eliminate a lot of tedious paperwork for you and reduce your stress load. Make employee timesheets that are easy to use, photocopy everything, and keep files organized by employee.

As for how you are going to pay your employees, I think semimonthly is much easier on your employee's cash flow than a once-a-month paycheck. You can choose to pay your farm employees by piece rate, by an hourly rate, or with a salary. Be sure that, whatever they are paid, they earn at least your state's minimum wage. If you pay by carton of berries picked, for example, the employee has to earn at least your state's minimum wage, even if she works slowly and doesn't pick that many cartons in an hour.

A word of caution, though, if you are thinking of paying by the hour: Hourly employees often try to get as many hours as possible and may take advantage of that situation. For example, they may collect and wash your eggs as slowly as possible or forget a tool back at your other ranch that they have to spend 30 minutes retrieving. You have to pay them no matter how slow, inefficient, or forgetful they may be. I remember watching one of our employees slowly walk from one side of our farm to the other, the distance of about half a mile, to turn on the irrigation pump. As he spent about 30 minutes making this ambling round trip, I thought to myself how much money that cost us. The employee could have hopped on our ATV and handled the irrigation in about five minutes, but instead this person was fully aware of what his slow walk was worth to him.

We also had another employee that sometimes would carelessly collect eggs into flats and usually break a dozen or two in the process. That carelessness just earned him an hour or two of wages while costing us a couple of dozen eggs plus his wages. He had no incentive to do a better job. When we switched our employees to monthly salaries with the option of profit sharing if our business did well, all of a sudden the incentive was there for them to work efficiently, because they no longer earned more money if they worked long hours. The potential of profit sharing also provided them the incentive of working to help our business succeed; almost immediately, those broken eggs became less and less of an issue.

The only major problem that transpired with this situation was that the employees were a little bitter that they were no longer earning as much as they possibly could by clocking more time. Racking up extra hours (and hence pay) had been their motivation to work. We did what we could to provide an interesting, varied, safe, and fairly paid working environment, but we eventually had to let these employees go because of our conflicting goals and work styles.

Now let's hear about a positive and successful example of a functioning family farm and healthy work environment. The Lazor family of Butterworks Farm excel at many things in running a thriving organic dairy business in northern Vermont and producing award-winning dairy products, but making family a priority and creating a respectful workplace with a low turnover is one area they are particularly good at. We even had the pleasure of visiting two other farms that got some of their training at Butterworks—Fair Food Farm and Misty Brook Farm—and they both had great things to say about their time working for the Lazors.

Scaling Up while Keeping True

BUTTERWORKS FARM, WESTFIELD, VERMONT

• Jack and Anne Lazor •

What first started out as a homestead for this couple of back-to-the-landers has turned into one of the most successful organic farmstead creameries in the nation. Jack and Anne Lazor took the money that her parents had saved for her to attend graduate school and instead invested it in 60 acres in the far northern reaches of Vermont. On a verdant, windswept ridgetop, where one almost feels compelled to belt out, "The hills are alive with the sound of music . . . ," the Lazors now raise around 100 Jersey cows for their certified organic dairy, the first one in the state of Vermont and most likely the entire nation.

JACK LAZOR
The Butterworks Farm grain mill is itself a work of art.

Jack and Anne met while both were working at Old Sturbridge Village in Massachusetts, a living history museum. They bought their initial 60 acres of raw land for $20,000 in 1976. They had a family milk cow, a large garden, chickens, and so on to provide their own food while they both worked off-farm jobs, Jack as a special education teacher in the local high school and Anne in mental health for the county. They built their own little modest house for an additional $12,000 and have since added to it. A few years later they were up to six cows and were processing milk on their kitchen stove to sell door to door, at the same time welcoming their baby daughter Christine into the world. The State of Vermont eventually learned about their unlicensed dairy and required them to get licensed, which they achieved in 1984.

The Lazors were among the first new-generation dairies in Vermont, indeed the country, that were milking and processing their own milk into various products. At this point it was mainly butter, cottage cheese, and some fluid milk. Jack and Anne were not following any sort of business plan at that point, just flying by the seat of their pants, as Jack told me. Even though the name of their farm is Butterworks, they stopped making butter because they never really had the right equipment and realized that selling the heavy cream was more lucrative than turning it all into butter.

Even though the Lazors had been farming organically since day one, they finally got certified in 1987 when they started wholesaling in earnest. The dairy license in the early years only allowed them to sell within the state of Vermont. However, they added their dairy products to a friend's Boston-area produce route and started pushing into other neighboring states without really realizing the regulatory implications. Demand for Butterworks cream and yogurt began to really pick up outside just the state of Vermont. In the winter of 1989–90 it seemed that each neighboring state's dairy inspector called them at the same time, requiring that they have a special interstate milk shipper's license, a regulatory hurdle that would require the costliest upgrades they would ever have to make. A wealthy businessman who heard of their predicament gave them a $100,000 loan with no terms, which allowed them to upgrade without taking on any bank debt (indeed, they may never need to pay that investor back).

The Lazors have shied away from debt financing for most of their 35 years' farming. If they do take on a loan, they only do short-term ones payable in no more than 90 days. They figure if they don't have the cash flow to support their upgrades, they won't be making them.

It seems that every large jump in scale for Butterworks Farm came because of pressure from regulators. Despite wanting to create a sort of underground economy, as many of the idealistic kids of the 1970s did, the Lazors had to temper their ideals over time and come into compliance with regulations. However, Jack is quick to point out that he thinks the Vermont Department of Agriculture, for the most part, wants to see more people farming and is not trying to grind people down with new regulations.

Until a few years ago, the Butterworks Farm business had tremendous cash flow. They were able to pay for most of their upgrades, such as a giant solar barn to comfortably house their cows in the winter, using only business proceeds. Things are a little different these days and profits seem to be a little tighter. Jack attributes that to a bit more competition in the marketplace, not only from other small-scale creameries getting into stores but also larger national outfits exhibiting power in the marketplace (like paying for shelf space and massive marketing campaigns, something Butterworks is not capable of doing), plus the current consumer fascination with Greek-style yogurt (a product that Butterworks does not make, at least not yet).

Other costs have gone up as the business grows; for instance, they now have hired a professional bookkeeper instead of having Anne do the books herself. Labor costs, especially payroll taxes and workers' compensation, have also gone up. Sometimes Jack feels as though he is in business just to keep his employees going. The minute you start to grow, you need more people; the more people the more complicated it gets and the bigger you have to grow to support all those people. This is certainly one of the trickier challenges of scaling up a business.

The Lazors recently made the decision to register their business as an LLC instead of the simpler sole proprietorship they had been operating under for most of their business's life. This will allow them to more easily transfer management and eventually ownership to the next generation (their daughter Christine and son-in-law Collin) but has also made life a bit more complicated and costly: little things such as having to transfer all of their vehicle registrations to the Butterworks Farm LLC instead of having them under Jack's name, or not being able to write checks anymore with the checkbook he used to keep in his back pocket. Now the bookkeeper writes the checks.

Over time they are setting things up so that Jack and Anne can step back from the day-to-day management. Anne used to be in the milk barn every day, checking on cow health, and she also did all the bookkeeping. Now she has more time for gardening, playing the cello, enjoying her grandchildren, and other pursuits such as community service. Jack still is in charge of getting all the crops planted and harvested, a job that keeps him intensely busy for the short summer season that Northern Vermont provides. Butterworks Farm grows all the forages and feed grains needed to feed the cows, something they are very proud of.

Currently, their enterprise consists of about 100 Jersey cows with anywhere from 43 to 50 cows milking at one time. They keep a closed herd, meaning they don't buy replacement heifers from off the farm—they raise their own. If a cow does not get pregnant for some reason, they usually process her into beef for their families to consume. If any cows die from old age or health problems, their carcasses are composted and returned to enrich the soil. Soil fertility is something Butterworks Farm takes very seriously. They use their own cereal grain straw as bedding for the cows in the milk parlor and in the wintertime barn. This carbonaceous material soaks up the urine and manure, then is removed to form long compost rows that get turned regularly. This compost goes back onto the fields to produce the forage and grains the cows need to fulfill their high energy requirements.

Unlike the more common manure slurry systems you see in Vermont, which have too high of levels of nitrogen and

SHAWN HENRY

Anne and Jack Lazor with a few farm friends

salt and actually degrade soil health when applied to the land, Jack's manure management system is one of the reasons he believes his farm is sustainable. The only things leaving the farm are a few minerals in the form of dairy products, which he can easily replace with rock dusts and limestone. His soil organic matter now hovers around 5 to 9%, twice what it was when he bought the land. Jack believes that one of the biggest sins of modern civilization is robbing carbon from the crust of the earth, an issue he is trying to redress by adding more carbon to the soils than he takes. Jack continues to be an avid student of all things soil related, from trace mineral fertilization and foliar feeding to increasing the nutrient density of foods. Stacks of books line every surface of his home to prove this point.

Butterworks Farm is now producing around five kinds of yogurt, two kinds of kefir (a drinkable yogurtlike product), buttermilk, heavy cream, and cheese. In addition, they can piggyback onto their extensive delivery route some of the grains they produce, including whole-wheat flour, pastry flour, cornmeal, spelt, and dry beans. Although grains are not a big moneymaker, by adding them to their current distribution route, they can at least keep their transportation costs down for these bulky products. The Lazors also sell some seed grains and animal feed right off the farm. They had to build their own granary and mill in the 1990s because that kind of infrastructure has essentially disappeared in the Northeast. If they had had to truck their grains all over to get them processed, it would have cost more than they were worth.

Jack got into grains because at first he just wanted to save his own seed and produce his own flour for baking bread. So he started out buying a used seed cleaner, then realized that if he could clean the seed for replanting his fields, maybe he could clean it well enough for human consumption, too. Often the economic equation doesn't work out, though, as he pointed out with an example: A pound of uncleaned oat seed can be sold for $0.40 a pound, and processed rolled oats for human consumption could maybe fetch $1.00 a pound. However, the six or so energy-intensive steps needed to create "rolled oats" would easily eat up that extra 60 cents you might earn. He is proud of being able to grow and save all his own seed, which reduces his dependency on the increasingly consolidated seed market. He can also sell his locally adapted seed to other local farmers who seek it out.

Jack Lazor has become the go-to person for grain growing and organic dairying, a fact that has kept him a very busy man. He is even working on a new book related to grain production in New England, and he sits on the board of the nascent Northern Grain Growers Association. Equally involved in leadership roles in the organic community, Anne Lazor served on the board of directors of The Cornucopia Institute for many years and is now on the board of the local Vermont group Green Mountain Farm-to-School.

Butterworks Farm uses around an 80% forage diet for their cows, and the rest is composed of organic grains and legumes grown on the farm. They are also collaborating with holistic livestock specialist Jerry Brunetti to develop a seed-sprouting system that would provide fresh, nutrient-dense sprouts to the Butterworks cows in the winter. During the grass-growing months, the milking herd goes out to fresh pasture between the day's two milkings. They stretch that as long as they can, usually until the end of October, after which the cows spend the winter in the giant, naturally lighted solar barn, where fresh straw or hardwood bark bedding is added daily. Their Jerseys look healthy and robust, with years of attention spent on breeding animals that are best adapted to their climate, produce high-quality milk, and enjoy vigorous health.

JACK LAZOR

Making high-quality compost from the farm's cow manure

JACK LAZOR

An integrated, self-sufficient model dairy farm

animals day in and day out. This level of care and mutual respect was heartening to see in a dairy. People treated well equals animals treated well. Happy animals equals happy employees. A self-perpetuating circle of positive reinforcement.

Other ways in which they are trying to lessen their environmental footprint include the installation of a behemoth 35kw wind turbine that powers 20 to 50% of their energy needs, depending on the season (winter wind is more common there). Likewise, they are installing a gargantuan wood-fired boiler that can be fed with scrap wood and lower-quality chunk wood instead of more common wood boilers that use processed wood chips—a product that is often 50% moisture and thus much less efficient. The boiler will provide high-pressure steam needed in their energy-intensive yogurt processing as well as provide building heat and pasteurization heat. All of the Butterworks Farm products come in recyclable packaging; Anne would like to see them go further by utilizing reusable glass someday instead of plastics.

The Lazors are keenly aware of some of the challenges that new farmers face—land access being the most prominent hurdle. Jack is glad to see more land trusts working to make their land available to beginning farmers. He thinks the local food market is still relatively small but growing every day. There is a large, untapped source of customers that shop in the mainstream grocery chains who

A recent marketing idea of Butterworks is to feature pictures and stories of each of their milking cows on their Facebook fan page. It's not too often that consumers get to connect an actual name and face to the cow providing their milk products.

How do they manage all of this? The first and second generation gets together at least once a week for a family meeting. Likewise, they gather all of their employees together each week for a staff meeting, actively soliciting their feedback. Many of their employees have been with them for years, and turnover is low. Jack attributes this to the simple philosophy of treating people how you would want to be treated. Although they can't offer full health insurance to their employees, they do offer to cover monthly medical expenses up to $300 a month if need be.

As I hunkered down in a corner of the milk barn one morning, I watched and listened to the employees' patient movements of the animals and kind words to them. Things like, "Hey, sweet girl," and "How are ya doing this morning?" was not what I expected to hear from a staff that has to milk the same would support a lot of new farmers if they could be lured over to the "other side"; namely, local food. Jack also recognizes that a lot of key infrastructure is missing in New England to support more farmers—things like slaughterhouses and grain mills, for example.

One bit of infrastructure that farmers perhaps could cooperatively purchase is a portable seed cleaner, an idea he learned from a grain-farming friend in North Dakota. Jack and Anne do their best to share their knowledge and experience with other new and established farmers in the region, and several of their former employees are now running farms of their own (the owners of both Fair Food Farm and Misty Brook Farm got some valuable training working at Butterworks).

Jack's advice to new farmers?

- I hope you have the farming bug. All your senses need to be open to learning at all times.
- Don't be daunted by the naysayers, but do ask questions, observe others, get more training.
- Be willing to work really hard. However . . .
- Make time for yourself, your health, and your family. The Lazors try to take a day off each week to do things such as sailboating with the family or swimming in a nearby swimming hole.

TAKE-HOME MESSAGES

- Be sure to clarify the goals, needs, and wants of other family members who will be directly or peripherally involved in the farming business.
- Create a transition plan for a multigenerational farm or one that you want to pass on to your children. Don't wait until a lawyer makes those decisions for you after your passing.
- Your spouse's off-farm income should not perpetually support your farming "habit." This is not the solution to marital bliss. This strategy might be understandable during the first five-year startup period, but not so much after that. Also, the working spouse needs to be on board for doing this—don't just assume it, though.
- Understand how the tax implications, liability implications, corporate structure, and so on might affect other family members in your family-farming operation. Don't blindside them with a huge tax bill or lawsuit—put practices into place to protect them.
- Don't assume family members will work for free; treat them with the same dignity and respect that you would all paid employees.

- At the very least try to provide workers' compensation insurance for all family members, employees, and volunteers. It will protect you in case they get injured on the job and will make your workers and volunteers more confident in their activities.
- You can provide a great work environment without breaking the bank. Above all, use respectful communication and constructive criticism and compliment good work.
- Don't be casual about hiring and firing. Follow the law, do your homework, ask for references, fill out proper paperwork, document hours, and so forth.
- If you have interns or apprentices, write up a formal contract; make all responsibilities and benefits clear; and understand what the law is in your state about payroll and apprentices. Don't use apprentices as a cheap form of labor; make sure you provide time off and educational opportunities and skill-building activities in addition to their vocational training (a.k.a. work).
- Understand what your county, state, and federal payroll obligations are. If you don't have time to fill out the paperwork and do the necessary deductions, pay a reputable payroll company to do that work for you. This is especially useful if you have more than a few employees.
- Try to make it a habit to have regular meetings with your employees and apprentices. Once a week at the beginning of the workweek is a good practice. Have someone take notes, and make notes available to all after the meeting (this can be as simple as posting handwritten notes up on a bulletin board). Always leave time in the meeting for employee feedback or discussion—that will make them feel respected and heard.
- Display the required employee laws on a bulletin board in a central location for all to read.
- Create a basic employee handbook so that everything is written and clear. Give a copy to all new employees, and make them sign that they have received a copy.
- Document all employee trainings, especially safety training, in a notebook or simple log. Have employees sign the logbook after each training they attend. This can protect you in case of a safety violation or an on-the-job injury or when getting an insurance review.
- Make safety a priority. Don't skimp on protective gear, give trainings, and don't be casual about operating heavy equipment or power tools.

- Conduct six-month or annual reviews with your employees. Provide raises, bonuses, equity options, or other rewards when appropriate for good performance.
- Provide opportunities for employee feedback and even encourage employee-led projects. This will incentivize independent ideas, foster problem-solving creativity, and provide "ownership" in the business.
- If you can't pay more in hourly pay or salary, consider offering profit sharing as an incentive for performance and for staying with the company.
- Don't ever verbally abuse your employees. Learn nonviolent communication, compassionate communication, or one of the other communication techniques for solving problems.

CHAPTER 13

ADD-ON ENTERPRISES AND VALUE-ADDED PRODUCTS

IT'S NOT EASY TO BE a profitable farmer these days, especially if the crops you grow or animals you raise don't earn you any handouts from the government. Your creativity, decision-making skills, and hard work are what you have to rely on. Sometimes you just can't sell enough bunches of chard, ears of corn, or sides of beef to make a proper living and fulfill your holistic goal. Since you've made it to the last chapter of my book, I am going to fill you in on a little secret. The money-earning potential of agriculture is not really in the basic production—it's in all of the added ways you can earn income off your land, your knowledge, and maybe the farm products of others, all of which I call "added value."

I'm not suggesting you all go out and build a carnival attraction on your farm or buy up the neighbor's cheap sweet corn, mark up the price, and call it your own. I want you to stick to your ideals of honesty, sustainability, and all the other values you have identified for your business and the way you want to live your life. But there is income to be made in the addition of strategic, complementary enterprises that are still related to agriculture. They just might be the key to the survival of your farm enterprise or to smoothing out your cash-flow stresses at certain times of the year. What might some of those added-value ideas be, and what are the important considerations for your decision-making process?

AGRITOURISM

A lot of trends are changing in the vacation and travel industry, especially as our economy changes, Americans work more hours, and fuel prices rise. This has given rise to the notion of "staycation," which essentially means taking a vacation within or nearby the region where you live, and maybe even coming home to sleep in your own bed for the night. Another simultaneous trend is rising, probably due to our increasingly urban nature and generational separation from the family farm, of people wanting to get out onto farms for pleasure, purchasing fresh food, and even learning something new.

These two trends combined provide ample opportunity for farms and ranches to provide opportunities for recreation, leisure, dining, and more on their land. From the more intense investment of creating a farmhouse bed-and-breakfast to simply having a weekend farm-stand or pick-your-own operation, you get to decide the intensity of the public's involvement. Here are some of the agritourism ideas I have seen and heard of:

- Farmhouse bed-and-breakfast
- Wall tents or yurts for sleepovers
- Monthly private dinner club or occasional farm dinners
- Private hunting/fishing club or stocked fishing pond
- Recreation: hiking, trail running, mountain biking, canoeing, birdwatching, horseback riding, cross-country skiing, and whatever else might be popular in your neck of the woods
- Farmstands and U-pick days
- Corn maze and pumpkin patches
- U-cut Christmas trees
- Weddings, parties, and other private celebrations
- Special events and classes, such as farmstead cheesemaking or gourmet cooking
- Yoga and meditation retreats
- Barn dances and hoedowns
- Kids' gardening and science camps
- Adult farming camps

The ideas are only limited by your creativity, access to capital, and probably a little local policy. Whether you choose to do things legally or not, you should at least call and find out what your current local land-use regulations will allow you to do and, within reason, extend to your neighboring property owners the courtesy and consideration that you would expect of them. Communicating your ideas early on to your neighbors might help prevent conflict later on. Here are some other important considerations to think of when planning your agritourism ventures:

- Road access and parking
- Signage
- Bathrooms
- Staffing needed
- Safety and liability
- Potential permits needed from your county or township
- Hidden costs, fees, taxes, and so on; for example, if you offer lodging, you may have to pay a local occupancy tax
- Will you need an approved commercial kitchen to do what you want to do?
- Income and expenses of the different enterprises as well as profit margins (include your agritourism enterprises in your strategic and business planning; refer to chapter 5)
- Compatibility with your privacy needs and your social or antisocial nature

In the farm narrative below, Skip and Erin of Green Gate Farms demonstrate some creative and fun ways to bring people out to their farm, not only adding profitability and improving year-round cash flow for their business but also connecting them to the community. Another great way to add value to your farming business is through classes, workshops, and camps. Green Gate Farms is doing more and more of these each year. Let's find out more about the concept of on-farm classes.

EDUCATION AND SKILL BUILDING

When my husband and I were raising livestock and poultry, we would often get requests from customers to come out to the farm to "kill their own" animal. The thought of someone who didn't know how to use a gun properly or who had bad aim when killing one of our pigs did not sit well in our minds. Nor did the thought of somebody with a dull knife trying to hack away at one of our chicken's necks. Likewise, did we really want that kind of liability of gun-toting, knife-wielding yahoos running around the farm?

Yet customers continued to ask, and it finally dawned on us that there was enough demand to turn this idea into a real moneymaker. So we offered our first pig slaughter and butchering class to a group of 10 people, most of them good customers we knew on a first-name basis. After the first class we asked for feedback, made a few changes, and offered four more pig classes that winter. Each pig class made us around 50% in net profits, or around $55 an hour for the time we had invested in it. Also, nearly everyone who attended one of our pig classes became a future customer of sides of pork, motivated to butcher at home once they had acquired the basic skills in our class.

In addition, my husband and I organized five chicken slaughter classes, which doubled as a way to "retire" our old laying hens. Other than some minor supplies such as plastic aprons and latex gloves, all of the class earnings were net profit because we used our spent layers as the "props." We made $85 an hour for those classes! Chicken class customers were then empowered to raise their own backyard birds and knew how to properly slaughter and eviscerate them. And they all continued to be great, loyal customers.

We conducted all of the slaughter classes in the winter, when our cash flow was normally quite limited; the classes made all the difference between being completely stressed out over money and not being stressed at all.

After this experience I think the opportunity for farms to deliver educational programs, do-it-yourself skills, and even leisure activities such as yoga classes is quite significant. However, if you have to hire all the "experts" to teach the classes, they may not be as lucrative. We hired a chef to help us teach the butchering portion of the pig classes, but the classes were still quite profitable for us, and the chef did well, too. That kept him coming back to teach all of the pig classes. Think about the expertise you may have: There are probably a few people out there who would love to learn what you know as second nature. You can monetize that expertise (sorry to sound so capitalistic) or use it to barter with others if you want to leave money out of the equation. Maybe you could teach a couple of chefs how to butcher a pig, and they could teach you how to improve your pork recipes and cooking skills.

Green Gate Farms of Austin, Texas, offers yoga classes, as well as kids' farm camps and even adult farm camps (for aspiring or wannabe adult farmers). The yoga classes are great to have on-farm because they encourage the farmers to stay limber—something they wouldn't have time for if they had to drive off the farm for lessons. The farm camps, while incredibly labor intensive and shifting time away from other farm chores, can build community, agricultural literacy, and income and create lifetime customers for the farm. You have to consider those kinds of trade-offs for your own situation. Read on to be inspired by Skip and Erin's story.

Farming for Community

GREEN GATE FARMS, AUSTIN, TEXAS

• Erin Flynn and Skip Connett •

What happens to a dream deferred? Fifty-year-old Skip Connett was not going to find out. Ever since he was a child growing up and working among the Mennonite farmers in Pennsylvania, he had a longing to have his hands in the dirt and to farm for a living. He even grew up on a gentleman's farm until he was 12, often raising some animals for the family and a giant garden that would sometimes produce enough abundance to sell a bit to the local grocer. However, his life took a turn toward writing, which he did for nearly 30 years, much of it in the field of public health at the Centers for Disease Control. His wife Erin Flynn was also a writer, working across the street from his CDC office at the American Cancer Society.

They watched the twin lifestyle diseases of diabetes and cancer (two-thirds of all cancers are lifestyle related, according to the CDC) escalate over the years and wanted to be more directly involved in prevention strategies. They would both get frustrated when these organizations would call for more research into the causes of obesity, diabetes, cancer, and other lifestyle diseases, when the causes seemed apparent. It's our diet, stupid. We need more local farms growing nutrient-dense food and getting it into the hands of everyone, regardless of income.

Skip and Erin had a comfortable life in Atlanta, both gainfully employed, owning their home, with three children between them to keep them entertained. But Skip could not get the farming dream out of his head. If he didn't start farming soon, he probably never would. How much would he end up regretting that later in life? He started pouring over books by Wendell Berry and Gene Logsdon, and eventually started freelance writing for the Rodale Institute. This allowed him to visit farms, attend conferences, and cautiously enter the world of sustainable agriculture without the risks of actually doing it.

Erin, on the other hand, wanted nothing to do with it. Despite volunteering for the organization Georgia Organics and her keen interest in healthy food, she had a historical understanding of the suffering of farmers, something she wanted no part of. Her family had ranched in Texas for generations, suffering from low prices, drought, pestilence, and even suicide. She was truly horrified by the thought of her and Skip trying to eke out an existence from agriculture.

She humored Skip for a while as he poured over land classifieds. His home state of Pennsylvania didn't look like an option because of high land prices, so they started looking in central Texas, where land was cheaper and they had family connections. Being closer to family and raising their young children in a farm setting was also appealing to them. After quitting their jobs, selling their home, and packing up their stuff, the deal on a ranch an hour outside of Austin fell through, forcing them to move into a rental home of Erin's mother. They both kept freelance writing as they started hunting around for other land opportunities. Skip got a job at a nearby organic farm, absorbing all he could from their small, intensive vegetable production. The couple heard about a rundown 5-acre parcel around 8 miles east of downtown Austin that was for rent. The historic old farmhouse, antique windmill, and weedy fields were surrounded by beat-up trailers, industrial parks, and an RV campground where people seemed to stay forever. Not exactly the neighborhood that would support an organic vegetable farm, but just maybe one that really needed it.

With a tiny bit of finances they gained from selling their Atlanta house they began farming with the name Green Gate Farms (named for their location—the east side is the agricultural gateway to Austin—and for the gate at the end of their driveway, which is nearly always open to the public). Right away they signed up 30 families for their first vegetable CSA, all of whom were willing to pick up from the farm. Skip and Erin added a little retail farmstand, too, first selling their own vegetables and over the years adding local fruit, milk, eggs, meat, honey, and other products from small farmers in their region. Each year they would scale up the CSA as they became more skilled at growing the vegetables and could take on more customers. They also added some laying hens for eggs, ducks, a few sheep and goats (mainly for entertainment), and some pigs. The couple raises a batch of feeder pigs each year for meat and also has a small herd of rare guinea hogs that provide venison-colored pork as well as glean the fields and provide fertility to them. Also, you can't beat their cuteness. If you want a pig that will delight your customers and roll over for belly rubs, this is the breed for you.

At first Erin had this romantic notion that living on a farm would give her time to write a book, do some sculpture, spend time with her children, maybe even go dancing at night. However, she is now 100% committed to the farming business, contributing to it equally—a factor that Skip believes has made them successful. Today, she writes about farming issues, contributing essays to local publications and writing the CSA newsletter and blog. But neither of them relies on writing jobs anymore to pay the bills. If it weren't for buying land to expand the farm two years ago, they imagine they would finally be farming in the black.

Since renting is never secure and their 5 acres on the edge of town has its share of issues, from high water rates to a disinterested landlord unwilling to provide a long-term lease, they decided to secure themselves a larger piece of land in the neighboring county. What is called the River Farm is 30 acres of mostly floodplain soils along the Colorado River (not the main Colorado, but a smaller river in Texas with the same name). Because they had been farming for less than 10 years, Erin and Skip qualified for the USDA Beginning Farmer loan, allowing them to purchase the land without much money down and low-interest rates.

Since cash flow is now tighter because of the mortgage payments, they are slowly making improvements to that farm as they have the money. Field crops that don't need as much daily attention, such as potatoes and garlic, are planted out there and sheep and goats have been added to control weeds and boost soil fertility. Fencing and a modest pole barn are progressing as their pocketbooks allow. Future plans include an agritourism program as well as housing for apprentices and "incubator farmers," who lease small plots of land.

Erin and Skip are realizing they can't farm forever, especially given their ages. They have identified very little support for sustainable and organic farmers in Texas, even though the demand is swiftly rising for their foods. Because of these factors, in 2011 they started a nonprofit called the New Farm Institute to help grow more organic farmers for Central Texas. In addition to summer camps and farming workshops open to the general public, they have started the first incubator farm in Texas. The first participants have been a twenty-something couple, followed by another young man, each given half an acre to try their hand at growing and marketing certified organic

HATTIE LINDSLEY

Bringing fun back to the farm through pony rides

vegetables. Erin and Skip have covered all costs out of their own pockets—providing access to equipment, greenhouse, tractor, and mentorship; the new farmers just pay for their own seeds and water.

Going forward, the New Farm Institute is seeking grants and partners to build the incubator program to enable it to take on more participants that graduate to increasingly larger plots. Especially important is funding for housing—already designed by local sustainable architects Stanley Studio—so new farmers can live and work on-site at their River Farm.

There are so many hurdles that farmers face, and there seem to be even more if you are within city limits. For a long time Green Gate Farms was charged for all of its irrigation water as if it were wastewater, even though that water was being used to irrigate food crops. The water was not sent down a sewage pipe for treatment, but that is what they were charged for, around $10,000 a year. Erin lobbied for years, even inviting city council staff and water regulators to their farm, until that law was finally changed.

Another impediment to farming in urban and suburban areas is how Texas calculates its property taxes. You don't qualify for an agricultural exemption until you have farmed the land for at least three years. Even then, some jurisdictions don't consider growing vegetables full-time to be farming: they want to see commodity crops or livestock on your land. They even prescribe a stocking density you must maintain to qualify for the exemption. Imagine if you were trying to practice good rotational grazing in an area with little rainfall and they told you to overstock your pastures. Micromanaging farmers doesn't really encourage any new folks to enter the field.

The other giant obstacle for farmers in Texas is the archaic water laws. It essentially boils down to "He who has the biggest straw gets the most water." A company can set up right next to your farm, drop an enormous well into the ground, and pull up as much water as possible until your well goes dry, and you have zero recourse (that's what happened to nearby Tecolote Farm, the longest-running CSA farm in Texas). For this and many other reasons, Erin and Skip helped found another nonprofit in 2010 called the Growers Alliance of Central Texas to help elevate farmers' voices in policy discussions and also to benefit from group purchasing of supplies. They aren't a cooperative yet but hope to be some day. The alliance holds field days on one another's farms followed with hearty potlucks full of laughter and conversation. The spirit of collaboration among seeming competitors is heartwarming and powerful.

CSA is at the core of Green Gate Farms' existence. Members are part of the farm; they are welcome to come out at any time to picnic, pet the animals, even go swimming or camping (at the River Farm). They also get discounts for the numerous events that Skip and Erin put on, from farm camps for kids and adults to yoga classes, cooking classes, and evening dances (or hootenannies, as they call them). For a long time all of the CSA boxes were picked up at the farm, which helped strengthen their customers' connection to the physical farm and also reduced the labor required

Collaboration instead of competition among Central Texas farmers

for distribution. However, to scale up their CSA and better compete with the more numerous produce delivery services creeping up, Green Gate Farms is setting up CSA drop-off spots all over Austin using volunteer community organizers.

Education is a huge component of the farm and an increasing source of revenues, too. Green Gate Farms provides training for hundreds of volunteers each year and offers several weeklong farm camps for kids during the year and shorter daylong farm camps for adults, too (especially relevant for adults interested in farming). They host school field trips, give pony rides, host weddings, organize farm dinners, and do fund-raisers as well. When the raging fires from Bastrop licked at their River Farm's edge during the most extensive drought in the history of the United States, Skip and Erin held a fund-raiser a month later to aid neighboring farmers who had lost everything.

Helping people is ingrained in their bones. They also fund-raise each year for their Sponsored Share program, which provides free CSA boxes—through the Dell Children's Medical Center's Texas Center for Prevention and Treatment of Childhood Obesity and the University of Texas Charter Elementary School—so underserved children can have access to good food and learn about healthier eating. The parents of those children receive free cooking classes using the contents of their CSA boxes so they can learn how to more effectively and deliciously incorporate a wider diversity of local vegetables into their diets.

Erin and Skip are continually trying to tie their food to public health efforts. They are not into farming because they are foodie gourmands, but rather because they want people of all incomes to have access to fresh, organically grown produce for their health. Erin wishes more farmers were invited to the table when decisions are made about food, land use, and public health. Sometimes, you just have to insert yourself into the solution, whether or not you were invited.

Erin and Skip work extremely hard, farm year-round, and are open to the public nearly every day. The prolonged drought, extreme summer heat, and wildfires of 2011 have set back their production, increased their operating costs, and exacted a large physical and mental toll on them. Green Gate Farms will be offering its first winter vegetable CSA and meat CSA this year to make up for some of the lost summer revenues. Yet Erin and Skip are trying to regain a little privacy and balance in their life, too. The gate now closes on Sundays, and they try to do very little by way of farm chores that day, instead focusing on their children and maybe even a little relaxation.

SKIP CONNETT

A Green Gate "hootenanny" builds social capital.

If they are burned out, however, you wouldn't know it by their smiles and enthusiasm for what they are doing. There is a missionary zeal to what they are doing—going into the new land and creating a new community around local food and farms.

Skip and Erin's advice for new farmers?

- Go slow!
- Farmers must diversify their incomes to ensure survival, but you need to be careful about taking on too many projects at once. Things tend to snowball on a farm.
- Also, be kind to yourselves and each other. Find a masseuse or acupuncturist who will trade food for health care. Make your family's physical, emotional, and spiritual health a priority.
- Lease first, buy later, so you can make mistakes on someone else's land and figure out what kind of farm you really want or need.
- Don't try to raise animals and vegetables at the same time unless you have experience in both.
- Realize that marketing requires just as much time and effort as farming if you want to be successful.
- If you do a CSA, overplant—better too much than too little.
- If you don't like people, don't do a CSA.
- To make a viable living, you need to extend your season, preferably to year-round.
- Attend conferences in your region, tour other farms, and make connections with other farmers, as there is strength and support in numbers. Don't isolate yourself.
- Don't despair about being small and "insignificant"—we are changing the food system one acre at a time.
- There are no quick fixes in organic farming, and there never are when you are working with nature.

VALUE-ADDED PRODUCTS

Value-added food products are those that take raw ingredients and change them through some process or addition of other ingredients, potentially making them more attractive to a consumer and worth more money. Value-added foods can often sit on a shelf or in the fridge or freezer for a while, potentially extending your sales season and allowing consumers to have local foods at different times in the year. There is a long list of advantages and disadvantages when it comes to value-added food products. I will share some that I have directly experienced or discovered when speaking with other producers.

VALUE-ADDED PRODUCT ADVANTAGES	VALUE-ADDED PRODUCT DISADVANTAGES
They generate sales when your fresh market foods are out of season.	There is a limit to how much processed foods people eat–they are only going to want so many jars of jam or hot sauce sitting in their fridge.
They keep your presence in the marketplace and space in the farmers' market.	Sales go down at farmers' markets in the winter–is it worth your time to still go with just value-added products?
You make more sales at existing markets (adds value for your labor).	Will customers' spending go down on your fresh stuff if they spend their money on your value-added products instead?
You can make use of lower-quality, blemished products.	Quality control is an issue if you utilize a copacker.
You can make use of lower-quality meat cuts or nonchoice grades.	Quality control is an issue if you utilize a copacker.
Ready-to-eat foods might attract new customers and get them tasting your foods (such as salami, jerky, dried fruit, roasted nuts).	
They allow you to sell a product out of season when price might be higher or market less saturated.	Volume may be much lower, though, so it may do little to even out cash flow or move product.
It keeps labor busy year-round.	Labor costs are a factor.
It provides work to a family member who may not want to work in the field or outside.	It can be labor intensive at times of the year when you may not have extra labor (such as during harvest).
It reduces market risk.	Insurances are needed; as for rules, regulations, permits, and licenses–get ready for lots of them.
It increases market diversity and options.	It is an investment into product that just sits there (with lower turnover); commercial kitchen and HACCP plans may be needed.

ETHNIC MARKETS

You might consider growing specialty crops, varieties, or certain animals to tap into particular ethnic markets. This isn't exactly an "added-value" market, but it does open up a new market sector and differentiates you from other producers of the same foods. I have to be honest that at first I was dubious that a farmer could cater to certain ethnicities and do so in a way that was profitable. A lot of immigrant populations (although certainly not all) are very price sensitive as they struggle to make ends meet in the United States, which is an expensive country to live in.

In my years of working at a small-farm business incubator in California, I worked with a couple of Latino farmers who tried to specialize in Mexican crops to sell to fellow Mexican immigrants. The first farmer focused on Mexican greens and potherbs and the other in Oaxacan staple crops. Neither farmer succeeded: the first because he could not command a high enough price to remain profitable and the second because of the same price issue and climatic differences. His beans, squash, and corn never matured in time—the wet, foggy fall ruined them all.

Because so many inexpensive foreign foods manage to travel the world and still arrive here cheaper than anything we can grow sustainably in the United States, price can be a big stumbling block when marketing ethnic foods. However, I think I have discovered a few ways to get around the price issue: Focus on freshness, quality, and harvesting the crop or animal as the culture or ethnicity would want it. The part about harvesting is critical.

For example, we discovered that raising particular breeds of chickens and processing them in a unique way would ensure sales among the Asian shoppers at our farmers' markets. We raised red- and black-feathered slow-growing broiler chickens, slaughtered them at smaller sizes the day before market, and kept the head and feet on per customer requests. Our chickens, even high-priced organic ones, would still sell out in the first 30 minutes of the market. Chinese, Vietnamese, and Korean customers would line up the moment we showed up. Our birds were so fresh that the only thing that would have made our customers happier would have been live birds that we killed for them on site! We probably would have done that if the markets had allowed it.

Renard Turner of Vanguard Ranch is one such farmer capitalizing on ethnic products, handily converting diverse American-born people and immigrants alike to his tasty goat meat products. Even though goat is the most widely consumed red meat on the planet, it has yet to really penetrate mainstream markets in the United States. Immigrants from Mexico, Africa, the Middle East, and other parts of the world gobble it up, though, and the Turners are doing their part to turn others on to this healthy and sustainable meat. They are also adding incredible value to the meat by converting it into ready-to-eat meals in their traveling commercial kitchen. They may add agritourism ventures, such as ethnic music and food festivals, to their farm as well.

Have Goat, Will Travel

VANGUARD RANCH LIMITED, GORDONSVILLE, VIRGINIA

• Renard and Chinette Turner •

As a military kid Renard Turner moved around a lot—not the kind of life that's conducive to witnessing the rhythms of farming, nor one that affords much of a chance to be involved in it. Nevertheless, in California, where his family settled for a few years, Renard got a chance to join his high school's Future Farmers of America (FFA) chapter, and he excelled in the agricultural science classes the school provided. He went on to work a couple of decades in the medical field but still held the dream of farming in the back of his mind.

After a time living in Washington, D.C., he and his wife decided to move to the country and found the verdant, rolling hills of central Virginia calling them. Gordonsville is ideally located within a two-hour drive of several large population centers—great for anybody trying to make a living direct-market farming, but far enough away not to just be an overpriced bedroom community full of unaffordable "ranchettes." The Turners started out as homesteaders, trying to be as self-sufficient as possible and even building their own house from scratch.

Back then, just as today, credit was tight for purchasing farmland, so they initially financed the farm purchase through owner financing. Both Renard and Chinette worked full time off the farm as they were getting things started. Chinette worked during the day, and Renard worked at night so he could spend his daylight hours getting the house and farm built. Chinette still works off the farm in the legal field, but the couple's plans to scale up the farm business a bit should afford her the opportunity to quit her day job. Renard is fully and happily

NATASHA BOWENS

The pride of ownership is evident at the front gate.

devoted to the farming business now. And boy, what a business it is. The Turners embody the notion of "agri-preneur."

They first dabbled in sheep and even ostriches but settled on meat goats about eight years ago. The forage base was not right for grass-loving sheep, and the ostriches were mean and simply not enjoyable to raise. Their land was a mixture of pasture and forest with quite a bit of browse and weeds—perfect for goats.

Right from the beginning, Renard knew that he had to direct-market his goat meat and get the highest value for it. He simply wasn't going to survive selling goats at auction or to a wholesaler. He stumbled upon the idea of a mobile concession trailer after a long conversation he had at the Virginia State Fair with a lemonade concessionaire. Renard learned that this lemonade man and his family spent half the year traveling to fairs and festivals and the other half of the year fishing in Florida—not a bad life. Renard figured the concept might work for him, too—spending part of the year traveling to fairs on the weekends and the other half of the year focusing on farm projects and production.

First, he got his goat production down to a science. He manages his pastures through rotational grazing, allowing the vegetation time to recover from the goat browsing and also for the parasites

NATASHA BOWENS

Renard Turner with one of his meat goats

to die off in the soil. He does not follow the "university recommended" goat stocking density, finding that a lower stocking density allows his vegetation better recovery time, lowers the parasite load, and reduces animal stress. As a result of rotational grazing and some selective planting of specific forage crops, Renard has to purchase very little hay for just a few winter months each year, and that amount is shrinking every year as his management improves. He is working with Virginia State University on an exciting new research project aimed at eliminating the need for purchased forages altogether.

Likewise, he does not grain-feed his goats at all. His antiparasital drugs just sit collecting dust in a medicine cabinet; through grazing management and breeding, Renard can virtually eliminate any parasite problems. In addition, a combination of electric fencing and livestock guard dogs (Akbash) prevents predators from bothering with his goats and also gives him peace of mind while he travels on weekends selling goat meat.

The Turners' goat breeding is now focused largely on Kiko and Valero maternal base (does) with Myotonic bucks. This combination is hardy, meaty, and muscular and produces great-tasting goat meat. The Turners raise around 160 kids a year for meat, slaughtering them at the fairly young age of six or seven months. Since they are not producing animals for maximum size to sell to wholesale meat markets, they focus more on taste and tenderness when they slaughter their animals. Feed efficiency and growth also starts to decline after goats are six months old anyway, and their land can only support so many animals without negatively impacting their forage base. They keep around 100 does and a few bucks for breeding.

Americans are a bit slow to pick up eating goat meat, even though it's consumed by the majority of people in the rest of the world. Still, Renard has no

RENARD TURNER

The Turners and their staff work the concession trailer.

trouble selling it; once people try a bite of one of the samples he often sets out, they inevitably end up buying a whole plate. Renard has perfected some delicious goat recipes that appeal to a broad swath of people, including goat burgers, goat kebabs, and curried goat. He travels to county and state fairs, food festivals, music festivals, and other events with his concession trailer. His wife often comes along and helps him cook, in addition to a couple of part-time employees who do most of the selling and money collection. Most meat producers' jaws are going to drop when I tell you this next tidbit. When I asked Renard what an average pound of his goat meat sells for after he transforms it into these value-added meals, he remarked, "Around $36 to $38 a pound, on the conservative side."

Buying and outfitting a food concession trailer, however, is neither cheap nor easy. Renard had to take food-handling classes, get his trailer licensed and inspected, and perfect his goat recipes so his meals would sell. Now the Turners are quite proficient at running the trailer, and the lines of eager customers form easily. Future ideas include organizing a solar-powered summer concert series at his farm to bring the customers to him instead of his having to travel so much. He also likes the idea of connecting more people to where their food is grown while entertaining them with good music from around the world. Each concert could have a different ethnic theme with goat recipes to match. He also dreams of adding an on-farm restaurant someday but knows that he can't do it all or he would risk spreading himself too thin.

Renard is emphatic that to be a good farmer these days you have to reinvent how you do business. A small or mid-scale producer can't compete on volume and prices typically, but you can compete on *quality*. If you don't love marketing, you can't do what he does. The Turners' innovations have been recognized through numerous awards, including "Innovative Farmer of the Year" at the 2010 National Goat Conference at Florida A & M University and 2011 Farmers of the Year from *Minority Landowner* magazine.

Renard and Chinette are also serious about sustainability; they built a passive solar geothermal home, grow organic gardens, and chose an animal enterprise with a small carbon footprint and one that suited their land the best. With support from the USDA Natural Resources Conservation Service cost-share program, the Turners recently installed a large hoophouse to grow organic vegetables year-round both for home consumption and for the ready-to-eat meals they sell out of their concession trailer. They also sell some of their excess vegetables to three food distribution hubs and a local grocery store. They have specialized in more obscure varieties and vegetables, utilizing locally adapted seed from Southern Exposure Seed Exchange. The Turners are crafting a life and a business that supports their love for the earth while feeding people healthy, delicious food.

Renard and Chinette's advice for new farmers?

- You have to have an entrepreneurial spirit; you have to control your marketing.
- Research and read as much as you can about the enterprises you have interest in. Then figure out a way to market those things. Part B (selling and marketing) is equally as important as Part A (growing and raising things).
- You have to be a master of many things—you have to be much more sophisticated these days to be a successful farmer.
- You have to become an expert at legal issues as well.
- Farming is not for the faint of heart—you have to be committed, self-motivated, and focused.

TAKE-HOME MESSAGES

- Don't open up your farm to visitors without addressing potential safety hazards.
- Don't assume that nobody will sue you because you are a farmer. Consider some general liability insurance before you allow

visitors on your farm. If you only have a couple of events a year with visitors, perhaps you can just purchase special-event riders on an existing insurance policy.

- Make it clear where people are to park, go to the bathroom, and wash their hands. If you are going to have a petting zoo or people touching animals, make sure they have a way to wash their hands afterward. You don't want anyone becoming infected with salmonella or *E. coli* after touching your animals.
- You don't have to be open all the time. You can still have privacy. Also, don't feel that you have to give a tour or stop what you are doing every time somebody wants to come for a visit. Create policies that work for you.
- Tourism projects can require a large infusion of capital and may not be economically viable. Do a marketing study and business plan for these enterprises just as you did for your farm-production enterprises.
- If you offer classes, workshops, and the like on your farm, there may be some local laws that affect what you do. For example, on-farm animal slaughter classes may be in conflict with animal slaughter rules in your county or state. In this case an anonymous phone call might obtain some information that could protect you from breaking the law. Make it clear that people are paying for the class, not the meat.
- You may also need to obtain special-event permits when you will be having a lot of people on your property for an event, especially if there will be music, traffic, and so forth generated by the event.
- A lot of farmers get requests to allow weddings on their property. This may sound lucrative at first, but you should do a thorough budget to find out if indeed you will make money by hosting weddings. It may require such large investments in insurance, permits, parking lots, bathrooms, kitchens, landscaping, and so on that it does not make sense.
- Value-added processing also involves considerable costs. You should do a thorough enterprise analysis to see if the profit potential is indeed there. There may be other reasons for doing value-added, such as extending your season, adding variety to your offerings, retaining customers over the winter, and reducing food waste, but those all have to balance out with covering your costs.
- Value-added processing is a completely different skill set, and it requires permits and certifications. Also, the time required for processing may come into conflict with your time for farming.

It might make sense to contract with a custom processor who already has significant experience making quality processed products instead of your adding it to your giant list of responsibilities.

- Selling to ethnic markets may be a great way to market certain animals, parts of animals, specialty crops, and so on. However, don't make assumptions about these or any other markets. Get direct feedback, pricing information, and so forth. Sometimes the logistical hassles, prices received, and other factors are not worth growing for these markets. For example, we could not sell live heritage-breed piglets to ethnic markets and adequately cover our costs—customers were not willing to pay for the heritage-breed premium. However, we could sell undersize feeder pigs to certain ethnicities looking for roasting-size pigs (they had to handle the slaughter themselves, but we could afford to sell them live).

  We could also sell most of our retired laying hens to Oaxacan families with pregnant women—the practice of making a soup from a fertile hen is an important part of their tradition. We let them capture the hens live and take care of the slaughter on their own; this freed up our labor because we didn't have to process the birds ourselves. Many farms we have visited are able to easily sell off roosters (the accidental kind that inevitably arrive when you purchase a batch of layer chicks) to immigrants living in the area. In that case it might make business sense to buy cheaper "straight-run" chicks that include hens and roosters, raise them up for a while until you can properly sex them, then sell off the live roosters. However, if you suspect that people are buying your roosters for cockfighting, you can have them kill the birds on the farm to prevent them from being used this way.
- There are many people of all nationalities that are looking for fresh, high-quality specialty foods—producing these foods might be a great addition for your farm's diversity. Want to add rabbits to graze and provide fertility to your orchard? Find a cluster of recent immigrants who are used to eating rabbit in their home country. Want to grow daikon radish to open up compacted clay soils and pull up deep nutrients? Look for Koreans who want to make their own kimchi, of which daikon is a key ingredient. The possibilities are only limited by your imagination.

References and Resources

⋆ Books ⋆

Advani, Asheesh. *Business Loans from Family & Friends*. Berkeley, CA: Nolo, 2009.

Berry, Wendell. *The Unsettling of America: Culture & Agriculture*. San Francisco: Sierra Club Books, 1977.

Blanchard, Chris, Paul Dietmann, and Craig Chase. *Fearless Farm Finances: Farm Financial Management Demystified*. Spring Valley, WI: Midwest Organic and Sustainable Education Service, 2012.

Kinsey, Neal, and Charles Walters. *Neal Kinsey's Hands-On Agronomy*. Austin, TX: Acres USA, 1999.

Logsdon, Gene. *Small-Scale Grain Raising, Second Edition: An Organic Guide to Growing, Processing, and Using Nutritious Whole Grains, for Home Gardeners and Local Farmers*. White River Junction, VT: Chelsea Green Publishing, 2009.

Ü, Elizabeth. *Raising Dough: The Complete Guide to Financing a Socially Responsible Food Business*. White River Junction, VT: Chelsea Green Publishing, 2013.

Wiswall, Richard. *The Organic Farmer's Business Handbook: A Complete Guide to Managing Finances, Crops, and Staff—and Making a Profit*. White River Junction, VT: Chelsea Green Publishing, 2009.

⋆ Magazines ⋆

Acres USA
4031-B Guadalupe Street
Austin, TX 78751
800-355-5313
www.acresusa.com/

Small Farmer's Journal
192 West Barclay Drive
Sisters, OR 97759
800-876-2893
smallfarmersjournal.com/

The Stockman Grass Farmer
234 W. School Street
Ridgeland, MS 39157
601-853-1861
800-748-9808
www.stockmangrassfarmer.com/

⋆ Internet ⋆

Agricultural Justice Project: http://www.agriculturaljusticeproject.org/
AgSquared crop planning program: http://www.agsquared.com/
American Grassfed Association: www.americangrassfed.org
Animal Welfare Approved program: www.animalwelfareapproved.org
Artisan Beef Institute (Carrie Oliver's consulting company): http://artisanbeefinstitute.com/
BluDog Organic Produce Services (Dina Izzo's consulting company): www.bludogops.com
California FarmLink: www.californiafarmlink.org
Carbonfree program: http://www.carbonfund.org/
Certified Naturally Grown: www.naturallygrown.org
CSAware management program: www.csaware.com
EPA's Agriculture Regulatory Matrix: http://www.epa.gov/agriculture/llaw.html
Finance for Food: www.financeforfood.com
Food Alliance: http://foodalliance.org/
Food Hub (a project of EcoTrust): www.food-hub.org
Indiegogo: www.indiegogo.com
International Farm Transition Network: http://www.farmtransition.org/
Kickstarter: www.kickstarter.com
LendingClub loan program: www.lendingclub.com
Local Dirt: www.localdirt.com
Local Harvest: www.localharvest.org
MarketMaker National Network: http://national.marketmaker.uiuc.edu/
National Young Farmers' Coalition: www.youngfarmers.org
New England Small Farm Institute: http://growingnewfarmers.com/
Northern Grain Growers Association: http://northerngraingrowers.org/
PickYourOwn.org website: www.pickyourown.org
Prosper loan program: www.prosper.com
Ranch Manager program: http://www.lionedge.com/
Real Time Farms: www.realtimefarms.com
Salmon-Safe program: www.salmonsafe.org
Small Business Development Centers: http://www.sba.gov/content/small-business-development-centers-sbdcs
Start2Farm: www.start2farm.gov

US Department of Agriculture
Farm Service Agency: www.fsa.usda.gov
Food Safety and Inspection Service: www.fsis.usda.gov
National Organic Program:
http://www.ams.usda.gov/AMSv1.0/nop
Natural Resources Conservation Service: www.nrcs.usda.gov
Rural Development: www.rurdev.usda.gov
US Patent and Trademark Office: http://www.uspto.gov/

⋆ Other Publications ⋆

Beckett, Jessica M. "Beyond the Discourse of Local Food: Land Trusts, Beginner Farmers, and Redefining Conservation in the Public Interest." (MSc thesis, University of California–Davis, 2011).

De Schutter, Olivier. *Agroecology and the Right to Food*. Report submitted by the United Nations Special Rapporteur, March 2011. http://www.srfood.org/index.php/en/component/content/article/1174-report-agroecology-and-the-right-to-food.

Fraser, Evan D. G., "Land tenure and agricultural management: Soil conservation on rented and owned fields in Southwest British Columbia," *Agriculture and Human Values* 21 (2004): 73–79.

Iowa State University Extension. *Improving Your Farm Lease Contract*, June 2011.

Organic Trade Association, "Industry Statistics and Projected Growth," June 2011; http://www.ota.com/organic/mt/business.html.

Index

About the Author

JENNIFER JONES

Rebecca Thistlethwaite runs a small-farm and food-business consulting firm called Sustain Consulting and is starting a small homestead in the Columbia River Gorge with her husband, Jim Dunlop, and their daughter, Fiona. They previously operated TLC Ranch in Watsonville, California, where they raised organic, pastured livestock and poultry for direct markets across Northern California. Rebecca currently serves on the board of the Sustainable Food Trade Association as well as the Columbia Gorge Earth Center and writes for blogs such as "Cooking Up a Story" and her own blog, "Honest Meat." Her other recent pursuits include training oxen, holistic financial planning, and learning how to make fermented foods. *Farms with a Future* is her first book.